DNA and Tissue Banking for Biodiversity and Conservation:

Theory, Practice and Uses

Published in association with IUCN – The World Conservation Union

Founded in 1948, The World Conservation Union brings together States, government agencies and a diverse range of non-governmental organisations in a unique world partnership: over 1000 members in all, spread across some 150 countries.

As a Union, IUCN seeks to influence, encourage and assist societies throughout the world to conserve the integrity and diversity of nature and to ensure that any use of natural resources is equitable and ecologically sustainable.

The World Conservation Union builds on the strengths of its members, networks and partners to enhance their capacity and to support global alliances to safeguard natural resources at local, regional and global levels.

DNA and Tissue Banking for Biodiversity and Conservation:

Theory, Practice and Uses

Edited by Vincent Savolainen, Martyn P. Powell, Kate Davis,
Gail Reeves and Angélique Corthals

Kew Publishing
Royal Botanic Gardens, Kew

PLANTS PEOPLE
POSSIBILITIES

First published in 2006 by
Royal Botanic Gardens, Kew
Richmond, Surrey, TW9 3AB, UK
www.kew.org

ISBN 1 84246 119 2

British Library Cataloguing in Publication Data
A catalogue record for this book is available from the British Library

Production Editor: Beth Lucas
Typesetting and page layout: Margaret Newman
Design by Media Resources, Information Services Department,
Royal Botanic Gardens, Kew

Printed in the United Kingdom by Cambridge Printing

For information or to purchase all Kew titles please visit
www.kewbooks.com or email publishing@kew.org

All proceeds go to support Kew's work in saving the world's plants for life

Contents

Foreword

Extraordinary changes have occurred with regard to the science, the policy and practicality of harvesting, storing and using DNA. Molecular techniques are now standard tools for phylogenetics and conservation biology, with the storage of tissue and DNA now a regular activity for biological collections. This timely and authoritative new work, a product of the UK government's Darwin Initiative and written by the leaders in the field, outlines these new opportunities and importantly provides the first practical and legal framework for these activities. These advances are paralleled by the maturing of conservation biology, a scientific and social discipline honed by the urgency of the expanding biodiversity crisis.

The potential of DNA technologies for understanding and documenting biodiversity was recognized in the 1990s; the subsequent decade has seen this technology move into everyday practice, where DNA work underpins taxon identification, conservation assessments and the genetic management of threatened taxa. The taxonomic and phylogenetic understanding of both cryptic and charismatic taxa is being constantly revised, whether tigers or pathogenic fungi. At the same time molecular tools are allowing us to marvel at the extraordinary pace and energy of evolution, as epitomized by the Succulent Karoo flora of South Africa in a recent article published in *Nature* by researchers at the University of Cape Town and Kirstenbosch. As more species decline in viability and as *ex situ* management becomes more closely tied to field conservation so these DNA and tissue archives will become invaluable resources. For example, at Fairchild Tropical Botanic Garden, with funding from the Institute of Museum and Library Services we have established a specialist palm and cycad DNA bank to allow us to undertake conservation research and to better manage that ephemeral resource, the botanic garden living collection. This DNA bank is one of the facilities of our Center for Tropical Plant Conservation with Florida International University.

This manual by Savolainen et al. brings together a comprehensive and practical review of the issues pertaining to DNA and tissue banking. DNA technology is already firmly established as a tool for implementing the Convention on Biological Diversity, its relevancy is proven in the crop genetic resource community and is being rapidly established for monitoring trade, biological inventory and in understanding the evolutionary landscape for conservation management. This manual shows the value of DNA and tissue banking, its application and importantly the legal and legislative context for the work. Research can no longer be undertaken in a legislative and legal vacuum.

Biological collections are both historical records and potent legacies for future use. It is exciting to review how our use of herbarium collections has developed since Luca Ghini made those first specimens 500 years ago and to then ponder how our DNA collections will be used in the coming decades and centuries. Indeed the science of DNA fingerprinting is only two decades old. However one thing is clear – if we are to see DNA technology contribute to biodiversity management and sustainable development we need to dramatically increase our investment in the scientific institutes of the developing world. This manual clearly moves DNA technology for conservation into the practical arena and will open the doors to both new practitioners and new applications.

Mike Maunder, PhD,
Director, Fairchild Tropical Botanic Garden and Chair of the IUCN-SSC
Plant Conservation Committee

Preface

Although our knowledge related to biodiversity science has increased exponentially, we are still losing biodiversity at an alarming rate. Assuming that no drastic changes in human behaviour occur, and given predicted climate changes, we can expect important modifications to the biosphere in the years to come. This will undoubtedly threaten the survival of our own species and may cause the extinction of a third of all species by 2050.

In this post-genomic era, our knowledge of the world's genetic resources holds many keys for preserving biodiversity and for the sustainable use of ecosystems by humankind; therefore DNA and tissue banking should be pursued with urgency. With the revolution in molecular biology and bioinformatics, models can now be used to assess ecological health, predict patterns of disturbance and devise conservation priorities and plans for species survival. Scientists have also taken up the challenges of assembling the 'tree of life', and within just a few decades we hope to discover the evolutionary relationships of all life on Earth. DNA barcoding technologies also hold the promise of democratising natural sciences, by enabling instant species identification and associated links with numerous sources of biodiversity data.

It is important that this research should be carried out legally, fairly and equitably, in accordance with national and international laws and conventions and in collaboration with the countries that hold this biodiversity. The tremendous benefits that may result from DNA-based research range from taxonomic information, conservation tools and exchange of technological know-how, to potential commercial products, and all scientists need to consider how they can effectively generate and share such benefits with the countries of origin.

Very few DNA banks have been established so far, and most zoological and botanical taxonomic collections are inadequate for long-term high-quality DNA preservation and extraction. This handbook aims to address this gap. It provides the necessary guidelines and encourages the establishment of DNA and tissue banks for biodiversity and conservation sciences, in addition to illustrating how such facilities can contribute to capacity-building in the developing world. There are four main sections to this handbook. Section A reviews the uses of molecular data in biodiversity and conservation, focusing on the latest developments in DNA technologies whilst also pointing out some of the limitations. Section B outlines a series of legal issues that are important for DNA banking, together with an explanation of the relevance of international conventions in conducting such activities. Section C describes practical considerations for DNA and tissue banking, ranging from health, safety and security issues, to laboratory set-up, database design and vouchering. Section D offers several case studies, from South Africa, North and South America, Australia and Europe, each focusing on differing aspects and practical applications of DNA and tissue banking in both plants and animals. Finally, eight appendices provide model legal documents and a DNA extraction protocol. We hope this handbook will find some space on the shelves of those interested in biodiversity and that it will prove useful to those involved in the conservation of our planet.

We are very grateful to Ingrid Nänni for her invaluable contribution throughout this project; to Nahem Shoa and David Davidson for their superb artwork; to Mark Chase for his comments on the manuscripts; to Beth Lucas for her assistance with the editorial work; to all of the authors for their contributions; and to the Darwin Initiative for funding this manual in part.

Vincent Savolainen, Martyn P. Powell, Kate Davis, Gail Reeves and Angélique Corthals (Editors)

List of Contributors

Kholiwe Balele — Leslie Hill Molecular Systematics Laboratory, Kirstenbosch Research Centre, South African National Biodiversity Institute, Private Bag X7, Claremont, Cape Town, 7735, South Africa

Stuart Cable — Millennium Seed Bank Project/Herbarium, Royal Botanic Gardens, Kew, Richmond, Surrey, TW9 3AB, UK

Mônica A. Cardoso — Laboratório de Biologia Molecular de Plantas – DIPEQ, Instituto de Pesquisas Jardim Botânico do Rio de Janeiro, Rua Pacheco Leão 915, Jardim Botânico, Rio de Janeiro, Brasil

Sérgio R. S. Cardoso — Laboratório de Biologia Molecular de Plantas – DIPEQ, Instituto de Pesquisas Jardim Botânico do Rio de Janeiro, Rua Pacheco Leão 915, Jardim Botânico, Rio de Janeiro, Brasil

Mark W. Chase — Molecular Systematics Section, Jodrell Laboratory, Royal Botanic Gardens, Kew, Richmond, Surrey, TW9 3DS, UK

Ferozah Conrad — Leslie Hill Molecular Systematics Laboratory, Kirstenbosch Research Centre, South African National Biodiversity Institute, Private Bag X7, Claremont, Cape Town, 7735, South Africa

Angélique Corthals — Ambrose Monell Collection for Molecular and Microbial Research, American Museum of Natural History, Central Park West, 79th Street, New York, NY 10024-5192, USA

Robyn S. Cowan — Genetics Section, Jodrell Laboratory, Royal Botanic Gardens, Kew, Richmond, Surrey, TW9 3DS, UK

Laszlo Csiba — Molecular Systematics Section, Jodrell Laboratory, Royal Botanic Gardens, Kew, Richmond, Surrey, TW9 3DS, UK

Kate Davis	Conventions and Policy Section, Royal Botanic Gardens, Kew, Richmond, Surrey, TW9 3AB, UK
John Donaldson	Kirstenbosch Research Centre, South African National Biodiversity Institute, Private Bag X7, Claremont, Cape Town, 7735, South Africa
Michael F. Fay	Genetics Section, Jodrell Laboratory, Royal Botanic Gardens, Kew, Richmond, Surrey, TW9 3DS, UK
Paulo C. G. Ferreira	Laboratório de Biologia Molecular de Plantas – DIPEQ, Instituto de Pesquisas Jardim Botânico do Rio de Janeiro, Rua Pacheco Leão 915, Jardim Botânico, Rio de Janeiro, Brasil
	Departamento de Bioquímica Médica, Universidade Federal do Rio de Janeiro, Rio de Janeiro 21944-970, Brasil
Félix Forest	Leslie Hill Molecular Systematics Laboratory, Kirstenbosch Research Centre, South African National Biodiversity Institute, Private Bag X7, Claremont, Cape Town, 7735, South Africa
	Department of Botany, University of Cape Town, Private Bag, Rondebosch, Cape Town, 7701, South Africa
Luciana O. Franco	Laboratório de Biologia Molecular de Plantas – DIPEQ, Instituto de Pesquisas Jardim Botânico do Rio de Janeiro, Rua Pacheco Leão 915, Jardim Botânico, Rio de Janeiro, Brasil
Timothy K. Fulcher	Molecular Systematics Section, Jodrell Laboratory, Royal Botanic Gardens, Kew, Richmond, Surrey, TW9 3DS, UK
Peter Goldblatt	Missouri Botanical Garden, P.O. Box 299, St Louis, MO 63166-0299, USA
Stephen Graham	Jodrell Laboratory, Royal Botanic Gardens, Kew, Richmond, Surrey, TW9 3DS, UK

Adriana S. Hemerly   Laboratório de Biologia Molecular de Plantas –
DIPEQ, Instituto de Pesquisas Jardim Botânico do
Rio de Janeiro, Rua Pacheco Leão 915, Jardim
Botânico, Rio de Janeiro, Brasil

Departamento de Bioquímica Médica, Universidade
Federal do Rio de Janeiro, Rio de Janeiro 21944-
970, Brasil

Edith Kapinos   Molecular Systematics Section, Jodrell Laboratory,
Royal Botanic Gardens, Kew, Richmond, Surrey,
TW9 3DS, UK

Siegfried Krauss   Kings Park and Botanic Garden, Botanic Gardens
and Parks Authority, West Perth 6005, Western
Australia

John C. Manning   Compton Herbarium, Kirstenbosch Research Centre,
South African National Biodiversity Institute,
Private Bag X7, Claremont, Cape Town, 7735,
South Africa

James S. Miller   Missouri Botanical Garden, P.O. Box 299, St Louis,
MO 63166-0299, USA

Alan Paton   Herbarium, Royal Botanic Gardens, Kew, Richmond,
Surrey, TW9 3AB, UK

Martyn P. Powell   Molecular Systematics Section, Jodrell Laboratory,
Royal Botanic Gardens, Kew, Richmond, Surrey,
TW9 3DS, UK

Domitilla Raimondo   Threatened Species Program, Kirstenbosch Research
Centre, South African National Biodiversity
Institute, Private Bag X7, Claremont, Cape Town,
7735, South Africa

Gail Reeves   Leslie Hill Molecular Systematics Laboratory,
Kirstenbosch Research Centre, South African
National Biodiversity Institute, Private Bag X7,
Claremont, Cape Town, 7735, South Africa

Vincent Savolainen   Molecular Systematics Section, Jodrell Laboratory,
Royal Botanic Gardens, Kew, Richmond, Surrey,
TW9 3DS, UK

Jill Sutcliffe   English Nature, Northminster House, Peterborough, PE1 1UA, UK

Ian Taylor   English Nature, Juniper House, Murley Moss, Oxenholme Road, Kendal, Cumbria, LA9 7RL, UK

China Williams   Conventions and Policy Section, Royal Botanic Gardens, Kew, Richmond, Surrey, TW9 3AB, UK

Maureen Wolfson   Research Services, South African National Biodiversity Institute, Private Bag X101, 0001 Pretoria, South Africa

The editors would like to acknowledge the support of all the institutions involved in the development of this book.

Section **A**

The DNA
Molecule and its
Uses in
Biodiversity and
Conservation

1. What DNA can – and cannot – be used for
V. Savolainen and M. W. Chase

The genome of an individual organism contains several thousand genes and many more spacers (inter- and intra-genes), all made up of DNA. DNA is a complex molecule composed of sugar, nitrogenous bases and phosphorous bonds, arranged in the famous double helix (Box 1). DNA can survive for several decades when dissolved in pure water or a dilute buffering solution, and it can also be preserved relatively well in dried plant or alcohol-pickled animal tissues, but DNA is sensitive to ultraviolet (UV) light and ionizing radiation as well as to many common enzymes and chemicals, including sweat and bleach.

When extracted DNA is held in a deep freeze DNA bank (-80°C), it remains useful for a series of experiments (e.g. DNA sequencing), but once irreversibly removed from living cells, DNA can no longer be used for the vast majority of biological applications and studies.

DNA archived in a DNA bank can be used principally for two main purposes. First, a minute amount of DNA can be used to amplify and sequence a specific portion of the genome. This means that many copies of that specific DNA region can be multiplied *in vitro* so that enough copies are detectable by DNA sequencing machines. The output is a text

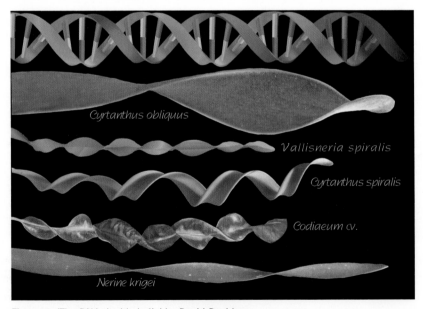

Figure 1. 'The DNA double helix' by David Davidson.

Box 1 The DNA double helix F. Conrad

It is now just over 50 years since Watson and Crick described the structure of deoxyribonucleic acid (DNA) in an article published in *Nature*. Rather than suggesting a three-chain structure as proposed by previous authors, Watson and Crick described DNA as 'two helical chains each coiled around the same axis'. They described the backbone of the double helix as consisting of alternating sugar (deoxyribose) and phosphate groups, with nitrogen bases joined to the sugar component. The bases are in turn held together by hydrogen bonds to stabilize the helix. They also found that only specific pairs of nitrogen bases can bond together: adenine (belonging to the purine chemical family, like caffeine and theophylline from coffee and tea, respectively) bonds with thymine (belonging to another family, the pyrimidines); guanine (another purine) bonds with cytosine (another pyrimidine). This led them to deduce that 'if only specific pairs of bases can be formed, it follows that if the sequence of bases on one chain is given, then the sequence on the other chain is automatically determined'. Hence the concept of the DNA double helix was born.

The term 'selfish DNA' was introduced by Doolittle and Sapienza in 1980 to account for the large proportion of non-coding DNA for which no functional significance could be found. In that year Orgel and Crick used the same term to propose that accumulation of non protein-coding DNA had little or no adaptive significance. These landmark discoveries and hypotheses have since given rise to a multitude of novel DNA-based technologies and techniques, which in turn have literally transformed our understanding of genes and organismal diversity and their associated evolutionary history.

In higher organisms the majority of genetic material is located in the nucleus and is inherited in a classical bi-parental Mendelian fashion. However, additional genetic information is compartmentalised into cellular organelles, comprising the mitochondrial and plastid genomes. Only plants possess a plastid genome, which is found in the amyloplasts, chloroplasts, chromoplasts and leucoplasts. Most often, organellar genomes are inherited via the maternal line; notable exceptions in plants include most conifers, in which the plastid genome is inherited paternally.

file containing the sequence of the nitrogenous bases (about 1–2,000) that make up the genetic region in question. Gene sequences can be compared to infer evolutionary relationships and build phylogenetic trees, that is, hypothetical organismal genealogies. Phylogenetic trees, whether based on the analysis of DNA sequences or on morphology, can be used to guide bioprospecting, discover areas of exceptional endemism and priority for

Figure 2. The myth of Jurassic Park. DNA has been successfully extracted and sequenced from museum specimens such as the woolly rhinoceros, quagga zebras, marsupial wolves and cave bears, but claims that more ancient DNA was sequenced, e.g. from dinosaurs, were later dismissed. (© Close Murray/Corbis Sygma)

conservation, identify factors promoting biodiversity and more. Phylogenetic trees based on DNA analyses are now more commonly used than those based solely on morphological characters because they are thought to be more accurate and reliable than trees based on morphological features alone.

Second, 'DNA fingerprinting' (or 'genetic fingerprinting') techniques can be applied to banked DNA. These techniques typically cut the DNA into smaller pieces and then, through amplification, selectively reveal banding patterns specific to species, populations, varieties or individuals, which can be considered equivalent to genetic barcodes. Fingerprints are used for identification or to infer genetic diversity and relatedness – information especially useful for conservation, breeding programmes, genealogical studies and genetic tracing. Small DNA sequences can also be used as DNA 'barcodes' for identification (see Chapter 3).

DNA is the basis of heredity, but it can only function as such when it is within living cells. Once DNA is banked, it becomes a non-functional unit of heredity; extracted DNA is 'dead'. The products of the genes held in DNA banks cannot be studied because the biochemical and cellular components necessary for gene function are not present in DNA extracts. Even more

Figure 3. Bee in amber, Dominican Republic. Although there have been attempts to extract DNA from insects preserved in amber, it is not possible to recreate living beings from such material (photo: Angélique Corthals).

speculative is the idea that banked DNA can be used to recreate the original organism from which it was extracted. Jurassic Park really is science fiction.

A DNA bank is a modern source of material that, together with herbaria and other museum collections, can effectively serve the sciences of biodiversity and conservation, and the sustainable use of that biodiversity.

2. Phylogenetics and conservation
F. Forest

Phylogenetics is the branch of biology that studies the relationships between organisms by constructing 'trees' (hypothetical sets of genealogical relationships) and examining the features that may explain the evolutionary patterns observed. The primary purpose of phylogenetic trees based on DNA sequence data is to establish relationships between taxa within a given group, assuming that relationships revealed by a gene or a combination of genes accurately reflect relationships between organisms.

2.1. Inferring evolutionary relationships for taxonomy

Reconstruction of phylogenetic trees to enhance our understanding of the evolutionary history of life has led to many significant and unexpected discoveries. During the past two decades there has been a tremendous increase in the number of published phylogenetic trees based on DNA sequence data (>4,000 in a decade). This is largely due to improved DNA sequencing techniques, such as advances in automated DNA sequencing and increased computer capacity that allows for the processing of larger data sets and application of complex mathematical algorithms. Many of these phylogenetic studies have revealed previously unsuspected relationships in all branches of the tree of life. In the plant kingdom, as a result of extensive sequencing of the plastid gene coding for the large subunit of RuBisCO (>10,000 *rbcL* sequences have been produced), phylogenetic analyses have led to a major re-classification of angiosperm genera and families by the Angiosperm Phylogeny Group (APG). One surprising case from this molecular phylogenetic study showed that the genus *Nelumbo* (water lotus) is not related to *Nymphaea* (water lilies) as previously thought, but rather to *Platanus* (plane trees) and proteas (Proteaceae). Similar re-classifications in the animal kingdom have also resulted from the use of DNA sequences to reconstruct phylogenetic trees. For example, a new large group Ecdysozoa (also called the 'moulting clade' because it includes animals with a cuticle that is shed) has been described and comprises arthropods and nematodes. This goes against the traditional belief that arthropods were related to other segmented animals, such as annelids. Likewise, elephants, hyraxes, sea cows, aardvarks and tenrecs, among others, previously dispersed in many different groups, have been put together in a group called Afrotheria, named as such because they are thought to have a common African origin. Similar examples of re-classifications are numerous and found in each part of the tree of life.

Figure 4. Scientists from around the world are collaborating to build the Tree of Life – the genealogy of all living and fossil organisms – which will become a vital resource for science and humankind. Here, David Davidson, a graphic designer in South Africa, who has a long association with SANBI and a passion for the natural world, has created a personal interpretation of this project. His baobab tree – a giant succulent (*Adansonia digitata*) – formed the centrepiece of the 2004 exhibit for SANBI at the Royal Horticultural Society Chelsea Flower Show, London, and was awarded a gold medal. This sequence of photographs shows the construction of the seven metre sculpture, which is now on permanent display at the Eden Project in Cornwall, UK.

Several methods are used for building phylogenetic trees, such as distance matrix methods, maximum parsimony optimality criteria, likelihood calculations and Bayesian statistics. Dozens of phylogenetic computer programs are available, but currently most phylogenetic trees are built with a limited number of these programs such as PAUP and PHYLIP for distances, parsimony and likelihood and MrBayes for Bayesian inference (see phylogeny program websites at http://evolution.gs.washington.edu/phylip/software.html).

2.2. Uses of phylogenetic trees: character evolution and dating

Unravelling relationships is not the only purpose for reconstructing phylogenetic trees. In fact, in more and more studies it is only the first step towards understanding the complex evolutionary processes behind the phylogenetic patterns depicted by a tree. Phylogenetic trees based on molecular sequence data can be used to examine and evaluate a multitude of evolutionary processes, through, for example, the reconstruction of ancestral morphological traits, estimation of divergence times between taxa, evolutionary correlation between traits, co-speciation, biogeographical patterns and speciation rates.

The mapping of characters, or traits, onto phylogenetic trees enables an estimation of the features that were found in the ancestor of two or more taxa. 'Maximum parsimony' is the criterion used most widely to infer the character states at the internal nodes of a tree. Reconstructions based on this criterion minimise the number of steps required to explain the character expressed in each terminal taxon. Likelihood calculations are increasingly used as an alternative for this purpose and several computer programs have already been designed to perform this task (e.g. Mesquite).

The next step in ancestral character reconstruction processes is to assess whether two characters are evolutionarily correlated. Comparative methods can be used to assess the correlation between two morphological traits, between a character and an attribute of the environment, or between any other combinations of features. Correlation of features on a phylogenetic tree can be accepted as evidence for adaptation.

For a long time, examining fossils was the only way of providing estimates for the appearance and extinction of various groups of organisms in the history of life. However, the fossil record is incomplete because not all groups of taxa are preserved in rock, and it is impossible to determine whether the oldest known fossil of a given group is actually the oldest taxon. The molecular-clock hypothesis, based on the assumption that DNA sequence data evolves at a constant rate through time, has opened up a wealth of possibilities for scientists to examine the timing of evolutionary events. However, it was rapidly shown that DNA substitution rates are not constant, but in fact vary through time and among lineages. In recent years many methods have been proposed to account for rate heterogeneity and make possible the use of DNA-based phylogenetic trees to assess the timing of evolutionary events. Whatever method is used to account for rate variation, the phylogenetic tree needs to be calibrated to transform relative estimates (branch lengths) into absolute ages (e.g. millions of years). The fossil record is often used for this task, although in some cases calibration can also be performed using geological or paleoclimatic data, such as the age of oceanic islands.

2.3. Measuring phylogenetic diversity to help prioritise conservation actions

Biodiversity is declining at an alarming rate in most areas of the globe as a direct result of human activities such as agriculture and urban development. The need to protect the remaining diversity of life on Earth is becoming increasingly urgent. Choices must be made about conservation of areas that contain crucial elements of biodiversity, and it is critical to determine which of these areas should be the focus of protection. Measures of diversity based on species richness and abundance are not sufficient for global conservation action, as species vary in their evolutionary histories and genetic distinctiveness. In addition to contributing towards understanding of relationships of life on our planet and how it evolved, phylogenetics can also be a powerful tool for addressing conservation issues. In recent years, several phylogenetic methods for evaluating biodiversity have been proposed and received increasing attention. These methods are aimed at identifying taxa and areas that should be prioritised for conservation.

The Phylogenetic Diversity (PD) method of Faith (1992) is probably one of the most promising of these and is based on a comparative summation of branch lengths between taxa. In other words, the phylogenetic diversity of a group of taxa is equal to the minimum sum of the lengths of all the branches that link these taxa together on a phylogenetic tree. The application of this technique in establishing conservation strategies is important, especially when based on molecular phylogenetic trees. For example, if a new biodiversity reserve is to be established and there is a decision to be made between two competing areas, the PD value of taxa present in each area could be used to determine which area has the highest phylogenetic value and should thus be designated as the reserve. Such a method should not be considered as the ultimate criterion in decision making, but it should be used as an additional parameter in conservation strategies in combination with economic, social and political considerations. Originally designed as a tool to understand the history of biodiversity, phylogenetics is now becoming an essential tool for the protection of its future.

3. DNA barcodes as tools for identification and tracking
M. W. Chase

A great deal has been written about the desirability of setting up facilities to produce 'DNA barcodes' for all life forms on Earth, but this section will not address these issues and instead will focus on practical issues relating to production of barcodes. Regardless of one's position on whether or not it is useful to have a universal system of DNA barcodes, there are undoubtedly benefits in moving in this direction, even if not all groups of organisms are represented initially.

A DNA barcode is a small section of DNA that has enough variation to make it possible to distinguish closely related species by analysing variability just within this region. In practical terms, it means that this piece of DNA has several base positions that differ among even closely related species. In animals, there appears to be one region, a gene in mitochondrial DNA called *CO1* or *cox1* (cytochrome oxidase 1), that is ideally suited for this purpose, but for plants the problem is more complex. It appears that at least two plastid DNA regions (or more) may need to be sequenced in plants because no single region alone has enough variation to serve this purpose. Usually plastid (in the case of plants) or mitochondrial (for animals) DNA is selected rather than the nuclear genome because few regions of the nuclear genome can be amplified (using polymerase chain reaction, PCR) with broadly applicable primers. Let us assume that some strategy will be worked out for all organisms that will enable us to discriminate between all species. The major problem to be faced in this endeavour is that not all species are easily discriminated; it is well known that in many groups there are 'cryptic species', that is, ones that we cannot identify based on morphology because they are too similar to others. In situations where there are more than two closely related species involved, systematists have used the term 'species complexes'. It is easy to see in such cases that adding DNA could help clarify the situation, but satisfactory resolution of these problematic species complexes will require careful studies of both morphological and molecular data. Let us also assume that such problematic groups will be studied with appropriate methods to determine which entities are real biological species and should be recognised.

There are many debates and a lot of work ahead if DNA barcoding is to be adopted as a general aspiration, but it is also clear that such methods will not make the situation worse. If accepted in a positive vein, DNA barcodes present the systematist with another tool that will shed some additional

Box 2 Identifying cryptic species V. Savolainen

Knowing the number and distribution of species on Earth is a key priority for the conservation of biodiversity. With the increasing availability and ease of use of molecular systematics tools, evolutionary history is now more directly incorporated into species delimitation, with a reported approximate 50% increase in species number resulting simply from applying phylogenetic species concepts (Agapow et al. 2004). Several species are not easily identifiable using morphological characters, although they are perfectly valid biological species, that is, discrete and genetically distinct. These 'hidden' taxa are called cryptic species, and they deserve full conservation attention, as do all other types of species.

DNA analysis is one of easiest and the most useful tools for discovering cryptic species. The scientific literature is full of reports of cryptic species, largely based on the examination of mitochondrial *cox1* gene sequences in animals, and ribosomal and plastid DNA in fungi and plants, respectively. In the last few years, scientists have discovered cryptic species of red algae, jellyfish, starfish, molluscs, crustaceans, fruit flies, ferns and even frogs, elephants and giraffes. Sometimes species have become isolated by geography but have remained morphologically identical, for example isopod crustaceans (*Mesamphisopus*) in

Figure 5. The African savannah elephant, *Loxodonta africana*, is one of the two species of elephant in Africa, with the African forest elephant, *Loxondonta cylotis*, recently discovered as a cryptic species. Genetic studies of elephant populations have shown that African forest and savannah elephants are as distantly related to each other as they are to the Asian elephant *Elephas maximus*; therefore African elephants do represent two separate species (and not one as previously thought), both of which deserve appropriate conservation management (photo: Peter Gasson).

Box 2 continued

different mountainous regions of the Western Cape of South Africa, or lamprey (*Lethenteron*) along a northern/southern disjunction in Japan. Sometimes long-term co-existence of cryptic species has occurred, such as fig–wasp associations where a surprisingly high proportion of cryptic species of wasp share the same fig hosts, thereby challenging the widely held belief of one species of pollinating wasp per fig species. In several cases, mechanisms that cause the formation of cryptic species can be complicated. For example, in some Atlantic bryozoans (*Bugula*), the presence of specific bacteria confers chemical resistance to predation and non-palatability to bryozoans located south of Cape Hatteras, and this might have initiated the reproductive isolation of these southern populations, which ultimately evolved into a separate cryptic species. In Indo–West Pacific urchins (*Echinometra*), divergence in sperm attachment proteins and sperm morphology have also produced distinct species, which are not identifiable by gross morphology alone.

Obtaining useful information from elusive species can prove difficult, and this is especially true for endangered and rare animals in which one might be able to rely only on field signs (e.g. hair, faeces, feathers). Fortunately, new DNA technologies, including the use of non-invasive sampling, can now help improve estimates of species delimitation and abundance, and centralized DNA bank facilities provide an obvious repository of genetic material from populations, races, varieties and species that can be used to discover cryptic species via DNA-based taxonomy and DNA barcoding.

light on the issue of species delimitation, and there are a great number of other useful things for which barcodes are well suited. Before DNA barcoding can be employed, there must first be a set of genetic standards established. This is the area that critics see as the most problematic, but once a set of such standards has been established, then unknowns can be analysed and compared with the set of standards to determine which species the unknown matches. There are many areas in which having such a system would be useful. Forensic workers would quickly be able to identify fragments (e.g. of plants) found on the scene of a crime, but in many such cases a simple species identification would not be helpful because many species are widespread and not restricted to just the areas in which the criminals were operating. For this tool to be more useful, knowledge of the population variation of the particular species would be required, but establishing a set of standards for this level of analysis would be a huge challenge and one that might be impractical at present – it could simply cost too much to be practical. Cultivated material presents an

Box 3 Wildlife forensics A. Corthals

Before the advent of DNA technologies, enforcement of wildlife laws mainly relied upon catching poachers in the act. Enforcement was made particularly difficult in situations where catching animals was legal only for some populations of a particular species, but not other populations of the same species, as has been the case for caviar and Caspian sea sturgeons, or whale meat and blubber. Wildlife forensics aims are the same as those of human forensics: to identify and convict offenders. Wildlife forensics involves sorting and identifying samples from the entire breadth of biodiversity, not only to detect crime, but also to aid other areas of wildlife management, such as conservation and preservation. DNA technologies now allow scientists to identify and distinguish species, populations and individuals using a range of techniques and DNA markers. This aspect intrinsically ties wildlife forensics to tissue and DNA banks and is often overlooked as an application of centralized genetic resources repositories.

Wildlife forensics has been particularly developed for the illegal harvest and trade, both domestic and international, in animals and animal parts. Illegal trade is one of the primary causes driving over-exploitation of threatened and endangered species, including food, traditional medicine (e.g. seahorses, shark oil, exotic plants

Figure 6. One of China's largest herbal medicine markets in the town of Anguo, Hebei province (photo: RBG Kew).

> **Box 3** continued
>
> and Chinese herbal medicine), pets (in particular birds), timber, animal furs and other products such as ivory, alligator skin, turtle shells and corals.
>
> A number of forensics laboratories have been established specifically to address wildlife genetics issues including, among others: (i) the US Fish and Wildlife National Wildlife Forensics Laboratory in Ashland, Oregon, (www.lab.fws.gov), (ii) the University of Maine Molecular Forensics Laboratory in Orono, Maine, USA (http://nature.umesci.maine.edu/forensics) and (iii) Trent University Wildlife Forensic DNA Laboratory in Ontario, Canada (www.forensicdna.ca). DNA banks and tissue storage facilities have played a major role in wildlife forensics and will continue to be at the forefront of conservation law enforcement.

even bigger problem than wild species because many horticultural cultivars are clonally produced and widely distributed.

The areas that would most benefit from a barcoding system would be those in which the group of organisms does not have much morphology (e.g. plant roots and seedlings, bacteria, unicellular eukaryotes and many worms and fungi) or in which fragments would help to establish that a particular species was present in an area even though no whole-organism sightings had been made (e.g. by comparing barcodes from a feather of a bird or a bone fragment). It would also be useful in cases in which different phases of an organism are not easily matched with the taxonomically understood phase (e.g. egg cases of molluscs or plant seedlings). Many species cannot be identified until they reach their reproductive phase, and in these cases barcodes would permit identification at any stage in the life of an organism.

Some people have suggested that a miniature genetic analyser would be useful for field identification of organisms, such as insects or plants, but unless the technology could result in inexpensive units (which might well be the case soon) there is little economic motivation for such a system – unless the animal or plant in question is a major pest, and then, generally speaking, its identity is easily established using more traditional methods of identification. Again, pathogenic fungi and bacteria would be more appropriate subjects because these are not readily identified by morphological traits. Curiosity about the name of a plant or animal does not provide the financial incentive needed to develop new DNA-based technologies. However, those technologies could be funded through those areas for which incentives do exist, and then once they become inexpensive enough it would

Figure 7. Science fiction? By 'crossing' the Star Trek Tricorder and mobile phone technology, a handheld field DNA barcoding device could be available in the near future to allow for instant identification of any living form on earth, and linkage with multiple sources of other related information. (© Collection Corbis Kipa)

become feasible to use them for other purposes. The tricorders of Star Trek come to mind: just point, press a button, and out comes the name and potential harm/benefit to be expected from that organism.

Tracking of tissue samples is another potential use of a DNA barcoding system, but as with the forensic uses discussed above, knowing that the sample is of species X may not provide the desired level of information, that is, establish which genotype of an organism is present. The knowledge that a given sample is from species X would not be specific enough to be useful. Barcodes would be useful for situations in which plants of a specific species, such as a rare or endangered species of orchid, were being sold, but in vegetative condition these could not be distinguished from other species common in cultivation. Some plant dealers might try to sell plants of the rare species labelled as the common species. A DNA barcode would enable inspectors to distinguish the plants and thereby discourage trade in the rare species. However, since most barcoding projects for plants propose to use plastid DNA markers, it will not be possible to determine that a plant is a hybrid or to discover both parental species of a hybrid because plastid DNA is maternally inherited. Other potential targets for tracking include those animal species being sold as 'bushmeat'.

Box 4 DNA barcodes versus genotyping V. Savolainen

DNA barcoding specifically refers to the identification of species, and although their set up requires the examination of DNA variation among populations, DNA barcodes are not intended to be identification tools at the population level. 'Genotyping', however, refers to the genetic characterisation of populations and thereby allows the inference of population history, origins and genealogies. Genotyping is performed using DNA fingerprinting techniques, which does not always involve sequencing of DNA fragments as is the case for barcodes. Several fingerprinting techniques produce experimental banding patterns (e.g. RAPD or AFLP techniques) that reveal the underlying genetic diversity of different populations (or even individuals), but they must not be confounded with DNA barcodes of species.

Like DNA barcoding, genotyping also has direct uses in forensics. For example, comprehensive fingerprints for different populations or individuals can be recorded and then used as templates to find the origin of an individual of unknown provenance. This technique has been used to track ivory tusks after extensive sampling of elephant populations. Similarly, genotyping of cycads in the wild has helped combat illegal collecting and can be used to distinguish between plantlets that have been propagated from nursery-registered versus wild parents.

B

Legal Issues Surrouding DNA Banking

4. DNA banking and the Convention on Biological Diversity
K. Davis, C. Williams and M. Wolfson

The Convention on Biological Diversity (CBD; see www.biodiv.org) has played a major role in shaping the political and practical climate in which biologists now conduct research. Anyone wishing to collect, exchange, study or sell plants, animals, other organisms, or their parts and derivatives (including DNA) needs to be aware of the broad scope of the CBD. DNA banks and tissue repositories may hold a vast range of material, from *in situ* and *ex situ* sources around the world. Through the biodiversity research that they enable, DNA banks have tremendous potential to help realise the objectives of the CBD, particularly working through CBD initiatives such as the Global Taxonomy Initiative and the Global Strategy for Plant Conservation.

4.1. Objectives and operation of the CBD

The CBD was opened for signatures by the international community during the Earth Summit at Rio de Janeiro, Brazil, in 1992. It came into force on 29 December 1993, after a sufficient number of countries (or 'Parties') had ratified it. As of February 2005, the CBD has 188 Parties (187 countries and the European Union). The CBD has three main objectives: (i) the conservation of biological diversity; (ii) the sustainable use of its components; and (iii) the fair and equitable sharing of the benefits arising out of the utilization of genetic resources.

The CBD definition of 'biological diversity' covers all levels of living organisms on Earth, from genetic diversity within species to diversity of ecosystems, including marine and other aquatic ecosystems. Within this broad mandate, the primary way the CBD attempts to stem the tide of biodiversity loss is through the 'ecosystem approach' to conservation. This is a strategy that promotes the integrated management of land, water and living resources to ensure conservation and sustainable use in an equitable way.

The core principle of the CBD is that States have sovereign rights over their biological resources, and have a corresponding responsibility to conserve and use these resources sustainably. However, recognising that the costs of conservation are greatest in countries that are rich in biodiversity, and that these countries tend to be poor in economic terms, the CBD aims to boost assistance for conservation in part through 'benefit-sharing'. Anyone wishing to gain access to 'genetic resources' is obliged to share any

Figure 8. River margin in Madagascar. The Convention on Biological Diversity covers the variability among living organisms from all sources, including diversity within species, between species and of ecosystems (photo: Stuart Cable).

resulting benefits (monetary and/or non-monetary) fairly and equitably with the countries of origin.

Unlike many other international instruments and conventions, the CBD is a framework convention. Rather than setting out clearly defined obligations or lists of activities, the CBD text establishes overall goals. Individual countries are then responsible for interpreting the provisions and decisions of the CBD according to their own national and regional priorities and capabilities, and implementing them through the development of national laws, strategies and other measures. This means that there is a great deal of variation between countries in how the CBD is implemented, which can make finding out about and keeping track of different countries' rules and regulations rather complex. To ease this process, each country is expected to nominate at least one National Focal Point (NFP), a person or agency that can provide information on national strategies, legislation and procedures that need to be followed to gain access to genetic resources.

The text of the CBD is set out in a series of Articles that outline the scope of the Convention and its organisational structure. Parties to the CBD meet at a biennial meeting (the Conference of the Parties, or 'COP') and agree on Decisions, which then become part of the CBD alongside the

Articles of the original text. The work of the CBD is further organised into thematic work programmes covering the Earth's major biomes. These programmes set out a vision for future work in these areas and provide a forum for experts in the different fields to decide on best practice (the most effective and ethical way to work). The COP has also identified important cross-cutting issues that need to be considered across all of the programmes, including taxonomy, plant conservation, access and benefit-sharing, public education, use of traditional knowledge and invasive alien species. Some issues with particular relevance to DNA banking are set out in Table 1.

In 2002, the COP adopted the 2010 Biodiversity Target: 'to achieve by 2010 a significant reduction of the current rate of biodiversity loss at the global, regional and national level as a contribution to poverty alleviation and to the benefit of all life on Earth', and this target was subsequently endorsed by the World Summit on Sustainable Development in Johannesburg later in 2002. In 2004, the COP adopted a framework to help assess progress towards 2010 and promote coherence between the many CBD work programmes, identifying focal areas and using indicators, goals and sub-targets. Research using DNA banks can contribute to several of the focal areas, such as 'the status and trends of the components of biological diversity'.

There is also the potential for 'protocols' to be developed to supplement the CBD. Protocols are separate legal instruments and only apply to the Parties that ratify them. The first such protocol to be developed under the CBD is the Cartagena Protocol on Biosafety, which covers issues relating to the transfer of living modified organisms between countries (see 4.11).

4.2. What are 'genetic resources' according to the CBD?

Article 2 of the CBD provides a list of definitions of important terms. 'Genetic resources' is defined as 'genetic material of actual or potential value'. 'Genetic material' is defined as 'any material of plant, animal, microbial or other origin containing functional units of heredity'.

The definition of genetic resources is important because it lies at the crux of many new strategies and laws governing access and use, and so has important implications for scientific research. However, the CBD definitions have given rise to different interpretations. On the one hand, 'potential value' is a broad term: who can predict the potential value of something in the future? On the other hand, 'containing functional units of heredity' narrows the definition: strictly speaking, this should include organisms and cells, but not DNA itself. DNA, although the unit of heredity, is not by itself functional. Some even consider that the term

Table 1	Some CBD cross-cutting issues of relevance to DNA bankers	
Issue	**Measures taken by the CBD**	**Description**
Access to genetic resources and benefit-sharing	Bonn Guidelines; new International Regime under negotiation	Provide guidance for governments developing laws and policies, and for users and providers developing policies, agreements and codes of conduct
Taxonomy	Global Taxonomy Initiative	Addresses lack of taxonomic knowledge and trained taxonomists
Plant conservation	Global Strategy for Plant Conservation	16 outcome-oriented targets with the long-term objective of halting the current loss of plant diversity by 2010
Sustainable use	Addis Ababa Guidelines	Describe general best practice across range of issues, e.g. stewardship, incentives, management, education
Technology transfer	Programme of work on Technology Transfer; Clearing House Mechanism (CHM)	Elements include technology assessment, information systems, capacity-building; CHM promotes and facilitates technical and scientific cooperation
Invasive alien species	Guiding Principles	Provide general guidance for governments on prevention, mitigation and control of invasive alien species

'genetic resources' refers more to the way that material is used (i.e. for genetic information and properties) than to the type of material.

In addition, countries have in some cases expanded on the definition used in the CBD in their national access legislation. For example, the African Model Law refers to access to 'biological resources' rather than genetic resources, and this term includes 'genetic resources, organisms or parts thereof ... with actual or potential use or value for humanity'. This definition easily includes herbarium specimens and DNA, whether functional or not. It is important for scientists to heed the relevant national legislation, and to bear in mind the importance of building mutual trust and partnership with in-country institutions and policymakers.

DNA is probably best described as a 'derivative' of genetic resources (although there is no agreed definition of derivative). The CBD itself does not specifically cover derivatives, but some national legislation does (e.g. that of South Africa), and it is possible that the new International Regime on access to genetic resources and benefit-sharing that is currently under negotiation (see 4.6) will apply to derivatives.

4.3. Article 15 on Access to Genetic Resources

Many national and regional governments have developed, or are developing, new laws and policies that set out how people may gain access to genetic resources in those countries and what benefits may be expected in return.

Article 15 of the CBD outlines the basic requirements for access to genetic resources and the subsequent obligation to share any benefits resulting from their utilization (Table 2). This Article has had a particular impact on scientists working in *ex situ* collections, as it affects how they acquire new material (through fieldwork and exchange programmes), use material in their research activities and exchange and supply the material to others. Article 15 recognises that States have sovereign rights over their natural resources, and so it is up to individual governments to determine how others may gain access to their country's genetic resources. It stresses that when designing new access procedures, States have an obligation to facilitate access for environmentally sound uses, that access should be with the prior informed consent of the Party providing the resources (unless otherwise determined by that country) and on mutually agreed terms. It states that these terms should outline the fair and equitable sharing of benefits arising from the use of the resources, and that research, where possible, should take place in, and with the full participation of, the provider country.

4.4. 'Pre-CBD' material

According to Article 15(3), the CBD's provisions apply only to material that is provided by countries of origin that are Parties to the Convention, and only to material that was obtained after the CBD came into force. The CBD, as with most international conventions, is not retroactive. Material acquired before the CBD came into force on 29 December 1993 is sometimes called 'pre-CBD material'.

Most material held by older *ex situ* collections falls into this category. Scientists and institutions need to consider how they will treat pre-CBD material, because it may affect their future relationships with institutions and governments in the countries of origin. It is increasingly regarded as

Table 2	Summary of CBD Articles addressing access and benefit-sharing
Articles	Themes
8 (j)	Benefits from the use of knowledge, innovations and practices of traditional communities should be shared equitably
15 (1)	National governments have authority to determine access to genetic resources
15 (2)	Access should be facilitated for environmentally sound uses
15 (3)	Genetic resources acquired pre-Convention are not covered by the CBD
15 (4)	Access to be on mutually agreed terms
15 (5)	Access to be subject to prior informed consent of the provider country, unless otherwise determined
15 (6)	Scientific research should be with full participation of/carried out in provider countries
15 (7)	Measures for fair and equitable benefit-sharing
16 (3)	Provider countries should be given access to and transfer of technology that makes use of their genetic resources
19 (1)	Provider countries should be enabled to participate effectively in biotechnological research activities
19 (2)	Provider countries should receive priority access to results and benefits from biotechnologies

best practice to treat pre-CBD and post-CBD material in similar ways as far as is possible, for instance, by committing to share benefits derived from the use of any material in a collection, regardless of the date of acquisition. It is not always possible to track the origin of older pre-CBD material, as a result of variable specimen data collection and ineffective tracking systems.

4.5. National implementation of Article 15

A growing number of countries have begun taking legislative steps to regulate access to their genetic resources and lay down measures to ensure fair and equitable benefit sharing. In general it has been the countries with the greatest biodiversity that have been first to implement these measures. As it is up to individual governments to decide on the precise details of how they regulate access, countries have developed a range of approaches that raise different expectations and have different requirements. In other words, access to genetic resources will differ from country to country, depending on national laws, policies and regulations.

Some countries have chosen to take a regional approach, believing that it makes little sense for one country to regulate access strictly whereas its neighbour, with similar fauna and flora, has little or no regulation in place. The countries in the Andean region (Bolivia, Colombia, Ecuador, Peru and Venezuela) have taken this approach: the Andean Pact Decision 391 of the Cartagena Agreement in 1996 establishes a common rule on access to genetic resources for all the member countries, although legislation is implemented at a country level. Another example is the African Model Law, which was developed by the Organisation of African Unity to provide a common standard for all African countries developing their national legislation on access to biological resources.

4.6. The Bonn Guidelines

The Bonn Guidelines on Access to Genetic Resources and Fair and Equitable Sharing of the Benefits Arising out of their Utilization were adopted by the COP in April 2002. They were designed by an expert panel to provide practical guidance to governments developing access and benefit-sharing laws and strategies and also to providers and users of genetic resources when they are negotiating contractual arrangements for access and benefit-sharing. The Bonn Guidelines outline in more detail what is expected to fulfil some of the key requirements in the CBD, such as seeking prior informed consent and negotiating mutually agreed terms. They also provide practical help for Parties and stakeholders, such as suggested elements for use in Material Transfer Agreements (MTAs) and a list of some of the monetary and non-monetary benefits that can be shared. When working with countries where there is little or no legislation on these areas, they provide a useful guide to accepted and current best practice.

The Bonn Guidelines are voluntary. Some countries now feel that too much emphasis has been put on the development of legislation by provider countries and not enough on ensuring that users are complying with access legislation. Following a call at the World Summit on Sustainable Development for an 'International Regime' covering benefit-sharing, negotiations are in process to develop this regime. It is likely to be based on the Bonn Guidelines but will also relate to other international instruments such as the International Treaty on Plant Genetic Resources for Food and Agriculture. Institutions with DNA banks should be mindful of new developments in the political arena.

4.7. Benefit-sharing

The CBD does not define benefit-sharing, although the Bonn Guidelines provide a useful indicative list of benefits that may be shared. It is important to remember that many of the benefits that result from scientific research and cooperation are non-monetary, though monetary benefits may sometimes emerge from commercially-oriented projects or fees for services.

Some of the benefits that may be generated around DNA banking include:

(i)   research-related benefits such as joint fieldwork and enrichment of national/local collections, joint research and co-authorship of publications, acknowledgment of sources of material in publications and databases, distribution of research reports and publications to colleagues in countries of origin, sharing of images and information in databases;

(ii) capacity building and education-related benefits such as sharing of technological expertise and know-how, provision of equipment, training workshops, higher educational opportunities and staff exchange; and

(iii) monetary benefits such as fees for research permits/use of staff/use of facilities, and in the case of commercially oriented projects, milestone payments, intellectual property rights and royalties.

4.8. Codes of conduct and other voluntary initiatives

A growing number of sectors and professional groups have developed voluntary initiatives and codes of conduct to guide the way they work while countries are going through differing processes of developing national legislation and building trust between potential partners and policymakers.

The Principles on Access to Genetic Resources and Benefit-Sharing were developed by an international group of botanic gardens and herbaria. They provide basic guidance on acquisition, use and supply of genetic resources, benefit-sharing and commercialisation. Institutions may endorse the voluntary one-page Principles and use them as a framework for an institutional CBD policy. A longer document, the Common Policy Guidelines, gives more detailed advice on implementation (see www.kew.org/conservation for more details, including an explanatory text).

The Micro-Organisms Sustainable use and Access regulation International Code of Conduct (MOSAICC) was developed by an international group of microbial culture collections. This voluntary code aims both to facilitate access to microbial genetic resources and guide development of appropriate MTAs. It provides an MTA checklist, setting out basic terms, use-specific terms and benefit-sharing terms (see www.belspo.be/bccm/mosaicc).

Both the Principles and MOSAICC are compliant with, and contributed to the development of, the Bonn Guidelines. Though developed by particular sectors, other sectors (such as universities and zoological collections) may find them useful for designing their own CBD-compliant policies.

Other examples include the Pew Conservation Scholars' Suggested Ethical Guidelines for Accessing and Exploring Biodiversity (1995) and the Food and Agriculture Organisation's Code of Conduct for Plant Germplasm Collecting, which has provisions on collector's permits, responsibilities of collectors, sponsors, curators and users as well as on reporting, monitoring and evaluating objectives (see www.fao.org/ag/agp/agps/pgr/icc/icce.htm).

Particularly in the absence of national law, some indigenous and local communities have begun to protect their rights over genetic resources and associated traditional knowledge, through declarations and by drawing up codes of conduct for those wishing to do research in areas in which they live. Examples include the Mataatua Declaration on Cultural and Intellectual Property Rights of Indigenous Peoples and the guidelines developed by the Inuit Tapirisat of Canada on Negotiating Research Relationships in the North. For more details and examples, see Laird (2002) and the CBD website (www.biodiv.org/programmes/socio-eco/traditional/instruments.asp).

4.9. The Global Taxonomy Initiative

Taxonomy is fundamental to the functioning of the CBD and measurement of its success. Absence of baseline taxonomic knowledge for many groups of organisms and lack of taxonomic expertise and infrastructure in many biodiverse countries are together recognised as the 'taxonomic impediment' to successful implementation of the CBD. The COP adopted the Global Taxonomy Initiative (GTI; www.biodiv.org/programmes/cross-cutting/taxonomy/) in 2002 to tackle this problem. The GTI aims to assess taxonomic needs and capacities, strengthen networks and infrastructures, facilitate systems for access to information and include taxonomic components where necessary in the thematic and cross-cutting work programmes of the CBD. Operational objectives have been developed to provide structure for Parties and regional groups to follow.

The GTI provides a valuable prompt to policymakers that taxonomy is important for decision making and that taxonomic research should be promoted and facilitated. DNA banks have a large part to play in enabling such research. However, there is no dedicated GTI funding from the CBD itself. To receive funding from the Global Environment Facility (GEF), the interim financial mechanism of the CBD, a taxonomic project must be put forward by a developing country Party (not a developed country institution) and be backed by the GEF and CBD National Focal Points, and taxonomy must be placed high on the list of that Party's priorities. There is, however, intense competition for GEF funds from all other areas of the CBD and other environmental conventions. Whether or not DNA bankers can win GEF funds by showing how they will help to implement the objectives of the GTI, they stand to gain political support for their work.

4.10. The Global Strategy for Plant Conservation

One of the most significant CBD initiatives for scientists working with plants is the Global Strategy for Plant Conservation (GSPC; www.biodiv. org/programmes/cross-cutting/plant/), adopted by COP6 in 2002. The ultimate and long-term objective of the GSPC is 'to halt the current and continuing loss of plant diversity'. The Strategy sets out 16 global outcome-oriented targets to be achieved by 2010, covering plant conservation, taxonomy, sustainable use, education and capacity-building. These targets are now being integrated into the relevant CBD work programmes. This target-led approach is proving popular with both policymakers and the botanical community. There are now clearly expressed measurable goals that can be tackled at different levels – global, regional, national and local. This focus should help to facilitate partnerships between those with different skills, knowledge and capacities.

DNA banks can enable biodiversity and conservation research that will contribute significantly to the GSPC targets. For example, work leading to identification and distribution of cryptic species will contribute to Target 1, 'a widely accessible working list of plant species, as a step towards a complete world flora', and Target 2, 'a preliminary assessment of the conservation status of all known plant species, at a national, regional and international level'. Research assessing and comparing genetic diversity of different areas can contribute to protected area planning and management and hence to Target 5, 'protection of 50% of the most important areas for plant diversity assured'. The development of DNA barcodes for eventual use by customs officers and other law-enforcement personnel may greatly assist with achieving Target 11, 'no species of wild flora endangered by international trade'. DNA banking has the potential to play a major role in many other areas of the GSPC. It is a good idea for plant DNA bankers to keep all CBD targets in mind when planning projects.

4.11. The Cartagena Protocol on Biosafety

Institutions and researchers that propose to use DNA held in DNA banks for projects involving genetic manipulation should be aware of the new Cartagena Protocol on Biosafety, which entered into force on 11 September 2003. The Protocol promotes the safe transfer, handling and use of 'living modified organisms' (LMOs) resulting from modern biotechnology that may have adverse effects on the conservation and sustainable use of biodiversity. It focuses on the movement of LMOs across international borders, and sets out requirements for 'advance informed agreement' in certain cases. The Protocol does not apply to LMOs that are used for pharmaceuticals for humans and addressed by other relevant international agreements.

The CBD defines an LMO as 'any living organism that possesses a novel combination of genetic material obtained through the use of modern biotechnology'. 'Modern biotechnology' is defined as 'the application of (a) *in vitro* nucleic acid techniques, including recombinant DNA and direct injection of nucleic acid into cells or organelles, or (b) fusion of cells beyond the taxonomic family, that overcome natural physiological reproductive or recombination barriers and that are not techniques used in traditional breeding and selection'. Further information and guidance can be found on the CBD website (www.biodiv.org/biosafety) and in the guide by Mackenzie and collaborators (2004).

4.12. The International Treaty on Plant Genetic Resources for Food and Agriculture

The CBD is not the only international instrument that covers access to genetic resources and benefit-sharing. The Food and Agriculture Organisation (FAO) International Treaty on Plant Genetic Resources for Food and Agriculture (the ITPGRFA, or 'IT'), which came into force on 29 June 2004, establishes a complementary regime covering plant genetic resources that are used and exchanged for food and agriculture. Unlike the CBD, the ITPGRFA has retrospective effect, so it will apply no matter when the material being transferred was acquired.

The ITPGRFA is designed to facilitate access between its Parties to 35 food and 29 forage crops. The crops are listed in an Annex, which can be amended with the consensus of the Parties. The ITPGRFA covers DNA bank material if the DNA being transferred is from listed taxa held by collections in the public domain and is being used for food- or animal feed-related research. Transfer of listed material will take place under a standard Material Transfer Agreement (MTA). This MTA is still being developed by the Parties, but once agreed, it will be non-negotiable and replace any in-house MTA for all ITPGRFA transfers. The MTA will bind recipients of these genetic resources and subsequent recipients to certain benefit-sharing arrangements. Recipients will not be able to apply for intellectual property rights (e.g. patents or plant breeder's rights) if this would restrict access to the material by others, unless they pay a mandatory fee (from the benefits arising from commercialisation) into the ITPGRFA's benefit-sharing mechanism, yet to be established. See www.fao.org/ag/cgrfa/itpgr.htm for details including the list of Parties and species in the Annex.

5. CITES and DNA banking
J. Donaldson

The Convention on International Trade in Endangered Species of Wild Fauna and Flora (CITES) entered into force on 1 July 1975 and was set up to protect species that are threatened by international trade. States that voluntarily agree to ratify the Convention (Parties) have to develop their own laws to make sure that CITES is implemented. As of 2005, there are 167 Parties to CITES and more than 30,000 species of animals and plants are protected by CITES regulations.

5.1. CITES Appendices

Species that are regulated by CITES are listed in one of three appendices:

Appendix I lists species with a high risk of extinction. CITES generally prohibits commercial international trade in wild-collected specimens of these species. However, trade may be allowed under exceptional circumstances, such as for scientific research. In these cases, trade requires an export permit from the Management Authority in the exporting country and an import permit from the Management Authority in the receiving country.

Appendix II lists species that are not necessarily threatened with extinction but that may become so unless trade is closely regulated. It also includes so-called 'look-alike species', that is, species of which the specimens in trade look like those of species listed for conservation reasons. Trade in wild-collected specimens of Appendix II species is allowed if the relevant authorities are satisfied that certain conditions are met. Above all, the Scientific Authority needs to be satisfied that wild harvesting will not have a negative effect on the survival of the species (in CITES language this is called a 'non-detriment finding'). International trade requires an export permit from the Management Authority in the exporting country. No import permit is required under CITES regulations, but some countries may implement stricter domestic measures that do require an import permit. For example, in the European Union, all Appendix II species are classified in either Annex A or B of the European Union Wildlife Trade Regulations and must be accompanied by an import permit from the relevant EU Member State. If material is sent to a country that implements CITES with 'stricter measures', both an export and an import permit may be required.

Appendix III is a list of species that have been included at the request of a Party that already regulates trade in the species and that needs the cooperation of other countries to prevent illegal exploitation. International trade in species listed in this Appendix is allowed only on presentation of the appropriate permits or certificates issued by the exporting country. For imports into the European Union, Appendix III species may be included in either Annex A or B, which require an import permit, or Annex C, which requires an import notification (Annex 2 of Commission Regulation 939/ 97, which is available from the competent authority in the EU Member State).

An up-to-date list of species included in the CITES appendices is available on the CITES website (www.cites.org). The lists are also printed in the CITES Handbook (last update 2003).

A list of Management Authorities is also available on the CITES website for checking whether species are covered by stricter domestic measures. The lists of species included in Annexes of the European Union Wildlife Trade Regulations are included in the full text of the regulations (http://europa.eu.int/comm/environment/cites/legislation_en.htm; also available via CITES websites run by individual EU Member States).

5.2. How is DNA banking affected by CITES?

In general, once a species is listed on a CITES Appendix, the agreement refers to all parts and derivatives (e.g. skins, teeth, leaves, roots, pollen, seeds) unless there is an annotation that specifically exempts certain parts or derivatives from CITES controls. For example, the annotations for many plant species listed in Appendix II exempt pollen and seeds from CITES controls. Unless there is such a specific exemption, the parts or tissues collected for DNA extraction from CITES listed species, as well as the extracted DNA samples, will fall under CITES controls. A proposal to exempt certain DNA samples, especially those derived *in vitro*, was rejected by the 13th Conference of the Parties held in 2004. As a result, the exchange of DNA samples between countries must comply with CITES regulations.

5.3. Exemptions for the exchange of scientific material between registered scientific institutions

Article VII of CITES makes provision for the exchange of CITES-listed material between scientific institutions as long as this exchange applies to the non-commercial loan, donation or exchange of herbarium specimens, other preserved, dried or embedded museum specimens or live plant material for scientific purposes. In these cases, the material must have a label attached that is issued or approved by the Management Authority.

Figure 9. Cycads (illustrated here by *Encephalartos latifrons*) are an example of species listed in Appendix I of CITES. This species is extremely rare in the wild, but large numbers of remaining plants have now been genetically fingerprinted. This information has shown that the living collection at Kirstenbosch Botanical Garden represents as much genetic diversity as there remains in the wild. This *ex situ* collection can therefore be targeted for breeding programmes and eventual *in situ* rehabilitation. DNA extracts from the plants used in this study are deposited in the South African DNA bank (photo: John Donaldson).

Box 5 A practical key to CITES J. Donaldson

When considering the acquisition or exchange of DNA material, the following guide can be used to determine what action is required to comply with CITES regulations.

1. Is the species listed in the CITES Appendices or in related lists for countries that apply stricter domestic measures (e.g. countries in the European Union)?

    Yes. Go to 2.

    No. The species is not under the control of CITES and no CITES permit is required.

2. (i) Is the material for non-commercial scientific use, **and**

    (ii) does it comprise preserved material or live plant material, **and**

    (iii) are both the exporter and importer registered with the CITES Secretariat as scientific institutions?

    Yes. The material qualifies for exemption under Article VII of CITES. The institution sending the material must affix a label that is issued or approved by the Management Authority in that country.

    No. The material requires a CITES permit. Go to 3. However, the institution may apply to their Management Authority for simplified permit procedures (as explained in 5.4).

3. Is the species listed on Appendix I, II, or III?

    Appendix I. An export permit is required from the country of origin (or a re-export permit if already in another country) together with an import permit from the country to which the material is being exported.

    Appendix II. An export permit is required from the country of origin (or a re-export permit if already in another country). For countries that do not apply stricter domestic measures, no import permit is required. For countries that do apply stricter domestic measures (e.g. countries in the European Union), an import permit may be required.

    Appendix III. A permit or certificate issued by the exporting country is required. For countries that apply stricter domestic measures, additional permits may be required (e.g. an import permit or import notification is required for countries in the European Union).

| Box 6 | A glossary of CITES terms and language | J. Donaldson |

Appendix	A list of species that are subject to the same trade restrictions and regulations. There are three CITES Appendices (I, II and III)
CITES Secretariat	The permanent staff who administer the implementation of the CITES Convention. The Secretariat is based in Geneva, Switzerland, and falls under the United Nations Environment Programme (UNEP)
Conference of the Parties (COP)	A meeting held every two to three years when delegates from all the Parties meet to vote on proposals for listing and de-listing and to agree on procedures and budgets
Management Authority	The institution designated by each Party to be the authority for issuing CITES permits, for enforcing CITES regulations, and for reporting to the CITES Secretariat on trade in species listed on CITES appendices
Non-detriment finding	An assessment from a Scientific Authority, which states that wild harvesting will not reduce the viability of wild populations. A non-detriment finding is required for any international trade in wild-harvested specimens of Appendix II species
Party	A country that has ratified the Convention
Range State	The country where a species occurs naturally
Scientific Authority	The institution designated by a Party to provide independent scientific assessments on the impact of trade on CITES listed species

In terms of current definitions, plant DNA samples qualify for exemption under Article VII. However, animal samples may not qualify for this exemption if they are not regarded as preserved specimens and if they contain any live tissues.

To qualify for exemption under Article VII, both the exporting and importing scientific institutions must be registered with the CITES Secretariat. The names of institutions that qualify for registration are submitted to the Secretariat by the relevant Management Authorities and a list of registered scientific institutions is available from the Secretariat or the CITES website (www.cites.org/common/reg/si/e-si-beg.shtml).

5.4. Simplified procedures for permits and certificates especially for animal samples

Although animal samples containing live material do not qualify for exemption from CITES regulations under Article VII, there are other mechanisms to simplify the issue of permits. Resolution 12.3, section XII, recommends that Parties use simplified procedures to facilitate trade (or exchange) that will have a negligible impact on the species concerned. In this case, the Management Authority must maintain a register of institutions that would benefit from simplified permit conditions. The Management Authority may then issue partially completed CITES permits to the relevant institution, which is authorised to complete the remaining information on the permit when material is exported.

5.5. Exchanging DNA samples between countries when one of the States is not a Party to CITES

If an exporting or importing country is not a Party to CITES, then material can be exchanged as long as the competent authority in that country issues a permit or certificate that substantially conforms to the requirements of CITES permits. These documents may be accepted in lieu of CITES permits by any Party.

6. Practical implementation of the CBD and CITES
K. Davis, C. Williams, M. Wolfson and J. Donaldson

This chapter provides practical guidance for DNA bankers on compliance with the Convention on Biological Diversity (CBD) and the Convention on International Trade in Endangered Species of Fauna and Flora (CITES). Details will inevitably vary according to national legislation, policies and regulations of the countries in which and with which your institution works and the resources and priorities of your institution.

6.1. Principles

Essentially, your institution will need to consider how it acquires, uses and transfers material:

(i) Acquisition – whether new material is obtained from the field (*in situ* conditions) or through exchange with other collections or individuals (*ex situ* sources), how will you ensure it is acquired according to national law, with appropriate 'prior informed consent' and under 'mutually agreed terms' and what benefits will be shared?

(ii) Use – how will the material (DNA, herbarium or museum specimens, living organisms, tissues, blood, cultures, other extracts) be used and curated by the DNA bank and related collections in the institution, and what benefits will be shared?

(iii) Transfer – can the material be loaned, exchanged, supplied or sold to others, and again, what benefits will be shared?

Prior informed consent (PIC) needs to be sought from the Party providing the material, unless otherwise determined by that Party (Article 15). This means seeking permission before carrying out activities, as well as providing information on how the material will be used, and possibly supplied, by the DNA bank and researchers. For instance, if someone else from your institution collected the material, then you will need to be sure that they had permission for DNA samples to be extracted. New PIC needs to be sought for any changes in use that are not within the scope of the original terms (e.g. commercialisation). Depending on national legislation, PIC may be needed from government, *ex situ* collection holders, private landowners, local communities and/or other stakeholders. Terms of access, use and benefit-sharing need to be agreed mutually by providers and users of the material. It is helpful if these 'mutually agreed terms' are set out in a written agreement, whether a simple permit or a more complex agreement.

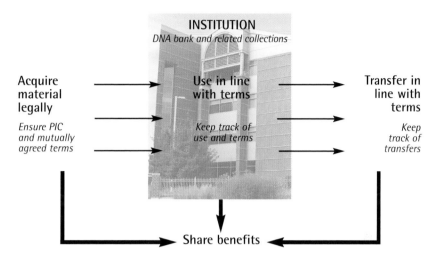

INSTITUTION
DNA bank and related collections

Acquire material legally

Ensure PIC and mutually agreed terms

Use in line with terms

Keep track of use and terms

Transfer in line with terms

Keep track of transfers

Share benefits

Figure 10. Basic principles for CBD implementation. Prior informed consent (PIC) and mutually agreed terms may be set out in permits, agreements and/or donation letters. Material Transfer Agreements (MTAs) are often used to set out terms for transfer/supply/exchange. If you wish to use or transfer material for purposes not covered by the original terms of acquisition (e.g. commercial development), you will need new PIC from the provider. Benefit-sharing should be considered at all stages.

It is also necessary to determine whether the material is derived from species that are listed on the CITES Appendices or are listed by countries that apply stricter domestic measures than those required by CITES, such as countries in the European Union. If so, then acquisition and exchange of material must comply with the regulations. You should check websites for the Management Authorities in both countries of import and export.

Developing and making publicly available a policy that sets out your institution's approach to CBD- and CITES-compliant acquisition, use, transfer and benefit-sharing will help to clarify institutional and staff responsibilities, as well as build trust with potential partner institutions, providers of genetic resources and policymakers.

6.2. Acquisition from *in situ* sources

If you intend to collect new material in the field, you will need to plan ahead and consider how, and from whom, you will obtain PIC and negotiate mutually agreed terms for your activities and research. The procedure and relevant stakeholders will vary depending on where you plan to work and which taxa and parts of organisms you are collecting.

Article 15(6) of the CBD emphasises that research based on genetic resources should be carried out with the participation of, and where possible in, provider countries. It is always a good idea, and increasingly a legal requirement, to work with an in-country partner. Setting up a collaborative partnership will enable more directed and constructive benefit-sharing. Partners can also help with researching and navigating the process of seeking and obtaining PIC, identifying other relevant stakeholders and complying with CITES procedures.

You will need to find out about relevant legislation governing access. In-country partners are likely to be aware of relevant regional, national and local legislation, although this information should be always be checked as there may be different procedures for in-country and external researchers. You can also consult the country's CBD National Focal Point (some countries have an NFP dedicated to access and benefit-sharing) or embassy. Colleagues with experience of working or collecting in the country are another valuable source of information, and you can also post questions on an internet listserve such as Permit-L (permit-l@si-listserv.si.edu).

Depending on where you are collecting or researching, you may need to get special permission from a government ministry, apply for a research and/or a collecting permit at national or local level, or find that you need to go through a more *ad hoc* process. If collecting on private or community land, you should get permission from private landowners or local communities. This could be in the form of a letter, but where literacy is an issue, this consent might be documented by audiotape or videotape. It is important to check through permits and other agreements to ensure that your institution can comply with the terms. Beware of collecting under a general permit of an in-country institution; this is unlikely to be considered as adequate PIC for your work as it may not cover the use, storage or exchange of material by your own institution.

In some cases, you may be required to develop a written agreement, such as a Memorandum of Understanding (MoU), a Material Transfer Agreement (MTA) or a benefit-sharing agreement, with the permitting agency or with in-country partners, setting out terms for acquisition, use, supply and benefit-sharing (see Appendices 2–5 and 7 for some examples).

To export material, you may need to apply for separate export permits, or in some cases you may need a written agreement such as an MTA. If you are intending to export or import living material, you will need to find out about relevant animal or plant health regulations (for instance you may need a phytosanitary certificate) or regulations on the transfer of micro-organisms. For CITES-listed taxa, you will need to apply for CITES permits (export permits and, in some cases, import permits) unless the material qualifies for exemption as an exchange between two registered scientific institutions. If it does qualify for this exemption, it is a good idea for the institution in the country of origin to send a letter to accompany the material confirming that it is donating the material in question. If you are not able to determine specimens of known CITES-listed groups to species level, it is possible to apply for permits at the generic level, with an explanation to the relevant Management Authority and letters of support from your partner institution.

Be realistic about the timescale required to contact partners and permitting authorities. If your project involves transferring large quantities of material, it is a good idea to contact customs officers in both countries in advance, so that the procedures are understood and the material can be cleared as rapidly as possible. Keep copies of all permits, letters of consent and related correspondence.

6.3. Acquisition from *ex situ* sources

Apart from fieldwork, institutions will regularly receive new material through exchange and donations from other institutions or individuals or by buying material from commercial sources. Similar principles to those applying when acquiring material from *in situ* conditions will also apply here. It is important to get some form of PIC from donors, to agree terms of transfer and to receive assurance that the material was originally collected legally and that the donors are entitled to pass it on to you.

Depending on national legislation, *ex situ* collections may be legally able to give their consent to supply material, although in some cases consent of a separate competent national authority may be needed. In some cases, the *ex situ* collection providing material will have a standard MTA that you can sign, which sets out terms of use, transfer to others and benefit-sharing.

If there is no such agreement, you could consider providing a 'donation letter' to the potential donor to check that the donor has acquired the material legally and they are entitled to supply it for your use, and to set out the terms under which you will use the material (see Appendix 6 for an example).

For CITES-listed taxa, material from *ex situ* sources must comply with CITES regulations in a similar way to material derived from *in situ* sources. If material is acquired from an institution that imported the original material under a CITES permit, then a re-export permit may be required in place of an export permit.

Ownership may be a thorny issue for institutions. Some collections will not accept material unless ownership is transferred to the collection, and some providers will not, or are legally unable to, transfer ownership. This issue needs to be handled on a case-by-case basis, but regardless of eventual ownership it is important for both parties to agree on the practical terms of use.

6.4. Use and curation

Once in an institution, material may be used in different ways by many different researchers. Your curation system needs to be able to make any restrictions obvious for all potential researchers (for example if material cannot be further sampled or transferred to others), and terms must stay associated with material at all times. The DNA bank database should have appropriate fields for any such restrictive terms, and because this information will also apply to specimen vouchers, restrictions should also be added to specimen databases and labels (see Chapters 9 and 10). Permit and/or agreement numbers could be added to databases and labels as applicable.

It may be helpful to prepare a 'use statement' for your institution: a document that sets out the general scope of research and common uses of material. This can be given to government authorities or potential donors when you are seeking PIC to acquire material. It will also help researchers to know what bottom-line terms are acceptable to their own institution so they do not risk getting material that then cannot be further used or curated.

If students and other scientific visitors use the material in your collection, you will need to ensure that they are aware of their responsibilities and any restrictive terms. If they wish to bring their own material, they should give you advance notice that it was legally collected and has appropriate permission for export (depending on the countries concerned they may need export, import, CITES and/or animal health or phytosanitary certificates). If they subsequently wish to donate the material to your collection, you need to be sure that they have permission to do so from their institution or, depending on national legislation, government authorities. The 'visitor policy' of your institution should outline these rules and responsibilities.

From time to time, researchers may need to check back to the original permits or transfer agreements for the material to see what uses were specified or restricted and who they might need to contact to ask for new permission for new uses. All these acquisition documents should be filed and linked to specimens so that researchers are able to find them.

6.5. Transfer of material

If your institution wishes to exchange and/or supply DNA material to other institutions for research purposes, you will need to set up criteria for dealing with requests. Some requests may come from individuals and commercial companies. Your institution should have a standard MTA that sets out the conditions under which material is supplied, for instance that material may be used for non-commercial purposes only (see Appendices 2–5). It is important to keep track of where DNA and accompanying MTAs have been sent, as you could be asked to provide this information at a later date.

When dealing with a request for the supply of material, you will need to check the terms of acquisition and consider whether (i) you are entitled to supply the material; (ii) there are any particular terms under which the material needs to be supplied (add these to the standard MTA); and (iii) the recipient is willing to sign your supply agreement.

Some DNA banks have decided to introduce a handling fee for this service, to cover the transaction costs. You will need to decide whether to do this, how much is reasonable to charge and how to handle requests for material from countries of origin.

6.6. Benefit-sharing

Institutions and individual researchers will also need to consider what kinds of benefits will arise from their work, how they can share them with countries of origin and how they can keep track of benefits that they have shared. In some cases, benefits are outlined in a partnership agreement, permits or even national legislation. In other cases, it will be up to your institution to ensure benefits are shared fairly for your general use of DNA material. Ensure that you maintain the link between the source of material in your collection and research that has resulted from its use. Information on countries of origin can, and should, be added into the collection details fields when you submit DNA sequences to databases such as GenBank (www.ncbi.nlm.nih.gov), EMBL (www.ebi.ac.uk/embl) or DDBJ (www.ddbj.nig.ac.jp), for material from *ex situ* sources as well as *in situ* sources.

When possible, it is useful to store information on databases about what benefits have been shared – for example, publications that have resulted from use of particular samples and where reprints or relevant publications have been sent. As the CBD and national laws develop, it is possible that there may in future be legal obligations to track material and the sharing of benefits.

Figure 11. Training and capacity-building are important elements of CBD implementation. Here, South African students learn the fundamentals of biotechnology for conservation at the DNA banking facility at the Kirstenbosch Research Centre, Cape Town (photo: SANBI).

6.7. Commercialisation

Although your DNA bank may not commercialise any material, you will probably receive requests to supply material for possible commercial research and so your institution should develop a position on use and supply of material for commercialisation. You will need to clarify the point at which you consider research to become 'commercial' with partners in countries of origin, as well as with third parties who request material from the bank.

You will need to get PIC and share benefits for any commercialisation of 'post-CBD' material (collected since the CBD came into force on 29 December 1993). You will also need to consider how your institution will deal with 'pre-CBD' material. From a purely curatorial point of view, it may be simpler to treat all material in your collection the same way, and this approach better follows the spirit of the CBD; however, you will not only need to consider your institutional capacity, but also accept that older specimens may not have full source data for countries of origin.

If your institution does not want to be involved in commercial research, another solution is to put those people requesting material for commercial research into direct contact with appropriate authorities and/or potential partners in the country of origin.

Whether or not your institution chooses to use or supply material for commercial purposes, it is useful to have a clear commercialisation policy. This should include (i) a clear and enforceable definition of commercialisation; (ii) a statement setting out the circumstances and terms under which the genetic resources may be commercialised; (iii) whether your institution will make a distinction between commercialisation of pre- and post-CBD material; and (iv) how benefits will be shared and the potential mechanisms for benefit-sharing (for example, milestone payments, a staff exchange programme and/or a benefit-sharing trust fund). You may want to cover other issues in such a policy, for example intellectual property rights, ownership, confidentiality and use of data.

One example of a commercialisation definition that you may wish to use is from the Principles on Access to Genetic Resources and Benefit-Sharing project (see the explanatory text at www.kew.org/conservation): 'applying for, obtaining or transferring intellectual property rights or other tangible or intangible rights by sale or licence or in any other manner; commencement of product development; conducting market research; seeking pre-market approval; and/or the sale of any resulting product'. Note that this is used as a definition of commercialisation *of genetic resources*, hence for example 'intellectual property rights' would include patents applied to the genetic resources but not copyrights on academic publications.

6.8. Intellectual property rights

Intellectual property rights (IPRs) are legal mechanisms designed to protect creations of the mind. They give exclusive rights of use, for a set period of time, to the holder or owner. Examples of IPRs include patents, which provide protection for new inventions, plant breeders' rights, which protect new plant varieties (called 'plant patents' in the USA) and copyrights, which protect original literary or artistic works (academic papers for instance). Although initial ownership of the IPR will usually be held by the original creator, IPRs may be sold, licensed, exchanged or given away, just as with physical property. In many cases, IPRs created in the course of employment will automatically belong to the employer. Many institutions may have already set up IPR policies, which should be consulted to clarify these issues.

Box 7 Written agreements K. Davis

There are several different kinds of written agreement that may be used by institutions to set out the terms and conditions under which biological material may be used and transferred, and how they will work with other organisations or individuals. Here we introduce two different types of 'Material Transfer Agreement': (1) those used for one-off transfers of material from your institution to others, and (2) those used for long-term collaborative partnerships between institutions that involve acquisition of material.

(1) An institution's standard agreement for one-off exchange/supply of material to other institutions or individuals is often termed a Material Transfer Agreement (MTA) or a Material Supply Agreement (MSA). This type of agreement sets out standard terms for use by recipients, e.g. further transfer, non-commercialisation and benefit-sharing. It may also be used to record certain additional stricter terms, on a case-by-case basis. This document should be written in clear language that is easily understandable to anyone wishing to use the material. Examples of institutions' standard MTAs/MSAs are provided in Appendices 2–5. These are provided as examples only and should be adjusted to conform to your country's national laws or for your institution's circumstances.

(2) A long-term collaborative agreement covering acquisition of material is an effective means of formally ensuring that, from the beginning of a project or partnership, both partners have a clear understanding, recorded in writing, of how PIC will be obtained from the relevant government authorities and other stakeholders, how the partners will work together and how any material collected may be used and curated. Sometimes this kind of agreement is called an Access and Benefit-sharing Agreement, a Benefit-sharing Agreement, or a Memorandum of Understanding (MoU), though the latter is a very broadly used term.

Such a collaborative agreement will usually be in addition to – not instead of – research, collecting, import and/or export permits, unless the partner in the country of origin of the material has governmental authority to provide full PIC for all these activities. CITES-listed taxa will always require appropriate CITES documentation in addition to any agreement.

Collaborative agreements must be set up on a case-by-case basis and will usually be drafted or checked over by a lawyer. Individual staff members will not usually have the authority or the expertise to negotiate this type

Box 7 continued

of agreement. Staff should be clear who at their institution has the authority to sign. This type of agreement may well take up to a year to draft, negotiate and finalise.

Copies of all such agreements should be kept where they can easily be located and consulted by members of staff, to ensure the agreement is being properly implemented.

Each collaborative partnership is likely to be different, depending on the various legal requirements of countries, administration of institutions and government bodies, existing infrastructure, areas of expertise and available funding; however, when setting up an agreement certain steps will be necessary in all cases:

   (i) Know your partners;

  (ii) Identify an institutional link – someone from each institution who can steer the agreement and project through the various stages and ensure that terms can be met;

 (iii) Find out about national laws – for example, by contacting the countries' national focal points on access and benefit-sharing and by working with partners;

  (iv) Identify relevant stakeholders and establish what PIC is required;

   (v) Define the areas of collaboration – especially for funded time-limited projects, it is a good idea to develop a project synopsis and attach this as an annex to the agreement;

  (vi) Agree the terms of use and further transfer of the material and benefits that can be shared (non-monetary and/or monetary);

 (vii) Check who can sign for each institution and/or government department;

(viii) Follow-up – ensure all partners keep track of material and promises.

Suggested elements for an agreement include:

   (i) Reason for collaboration;

  (ii) Agreement to work within CBD, CITES, national laws;

 (iii) Definitions;

  (iv) Areas of cooperation;

   (v) Permission to collect material;

  (vi) Where material will be kept;

> **Box 7** continued
>
> (vii)   How the material may be used;
>
> (viii) Benefit-sharing;
>
> (ix)   Non-commercialisation;
>
>   (x)   Transfer to third parties;
>
> (xi)   Use of data;
>
> (xii) Legal clauses such as duration, renewal, termination.
>
> Appendix 1 of the Bonn Guidelines (see Chapter 4) provides a fuller list of suggested elements for MTAs. The MoU covering the Darwin Initiative DNA Bank project between RBG Kew and the South African National Biodiversity Institute is provided in Appendix 7. For further examples of how some of the legal clauses may be worded, see the model Material Acquisition Agreement in Annex 2 of the Principles on Access to Genetic Resources and Benefit-sharing explanatory text at www.kew.org/conservation and the model framework for an Access and Benefit-Sharing Agreement in Cheyne (2004).

There is widespread political and ethical concern about the patenting of gene sequences. Although it is unlikely that such IPR issues will arise in the course of basic biodiversity conservation and research, all DNA bankers should be aware of this concern. This is a clear case of commercialisation of genetic resources, and any restrictions on commercialisation should be made clear at the time of acquisition and when transferring material under MTAs. When a decision is made to apply for a patent, it is good practice, and in some jurisdictions a legal requirement, to disclose the country of origin of the material and provide evidence that you have obtained PIC for commercialisation.

In the context of DNA banking for biodiversity research, IPR issues will usually concern the ownership of intellectual work done on material in the collection, rather than patenting of gene sequences: for example, academic papers written, databases created and information contained in databases. Often several people or institutions will have been involved, and you will need to decide on co-authorship and how to share ownership of the IPR. In addition, IPR issues may arise from contractual obligations you have entered into (e.g. transfer of copyrights to publishers).

In light of how data are now used, and particularly with the increase in use of online databases, it is a good idea to fully consider your institution's position on IPRs, and to include necessary information on how you wish copyright to be assigned in documents such as MTAs as well as in data use agreements or database login pages.

Section C

Practical Considerations for DNA and Tissue Banking

7. DNA extraction protocols
L. Csiba and M. P. Powell

Although there are many protocols for extracting DNA, they all follow the same fundamental principles. In this chapter, the basic steps for total genomic DNA extraction from plants are outlined; a more detailed extraction protocol is provided in Appendix 1.

Samples processed following the protocol described are theoretically stable for an indefinite time period when stored at -80°C. Most of them are stable for at least a few days at room temperature as well.

Various types of material can be used for extraction, including seedlings, leaves, flowers, cotyledons, seeds, endosperm, embryos, tissue culture callus and pollen. In addition, milligram amounts of tissue can be used when sample size is limiting. The starting material can be fresh, silica-dried, frozen or from herbarium specimens.

After collection of the material to be extracted, the first step in DNA extraction is to homogenize the tissue, rupturing the cells and 'isolating' the nuclei, mitochondria and plastids from which the DNA is extracted. This is normally achieved with use of a buffered salt solution containing a detergent (most often cetyl trimethyl ammonium bromide, CTAB), which serves to lyse nuclei, mitochondria and plastids and release DNA. The detergent also inhibits any nuclease activity present in the preparation. It is important to grind the tissue thoroughly, although this procedure needs to be performed rapidly to prevent DNA degradation.

The next stage of the extraction process is designed to purify the extract. The goal of all subsequent purification steps is to separate the DNA from contaminants, such as proteins. This is typically achieved by extracting with an active protein denaturant, such as 'SEVAG' (chloroform:isoamyl alcohol 24:1) which causes proteins in the preparation to precipitate from solution. The suspension is then simply centrifuged to separate two distinct phases. After centrifugation, two phases should be evident, with DNA in salt solution in the upper aqueous phase and polysaccharides in chloroform solution in the lower phase. A 'boundary' layer will be evident between the two phases, formed by protein precipitate. The aqueous phase is then pipetted away from the SEVAG and boundary regions.

The nucleic acids are then precipitated from the salt solution by the addition of cold ethanol or isopropanol. The samples are then spun in a clinical centrifuge to collect the precipitate; the supernatant is poured off

Table 3	Typical equipment and consumables required to extract DNA using the 2× CTAB procedure of Doyle and Doyle (1987)

Equipment	Labware	Consumables	Chemicals*
Autoclave	Beakers (2 L, 1 L, 500 ml, 100 ml)	Autoclave tape	2-mercaptoethanol (100 ml)
Balance (sensitivity 0.01 g)	Falcon tubes (50 ml)	Dialysis tubing (6.3 mm × 30 m)	Agarose (500 g)
Centrifuge, 24 position standard rotor for 1.5 ml tubes	Forceps	Labelling tape	Bromophenol blue (loading dye) (5 g)
	Full-face visor	Microcentrifuge tubes (1.5 ml)	Buta-1-ol (2.5 L)
Centrifuge, fixed angle rotor with adaptors for 50 ml tubes	Green screw cap tubes (12 ml)	Non-sterile gloves	Caesium chloride (100 g)
	ISO bottles graduated for stock solutions (2 L, 1 L, 500 ml)	Parafilm	Chloroform (2.5 L)
Electrophoresis documentation and analysis system		Particulate respirator (face-mask)	Ethanol 99.7–100% (2.5 L)
	Measuring cylinders (2 L, 1 L, 500 ml, 100 ml)		Ethidium bromide (30 ml)
Electrophoresis power supply		Pipette tips	Ethylene diamine tetra-acetic acid (EDTA) (500 g)
	Mediclips	Weighing boats (100 ml)	
-20°C Freezer	Mortars and pestles		Isoamyl alcohol (1 L)
-80°C Freezer		Ziplock bags (80 mm × 100 mm)	Isopropanol (2.5 L)
	pH meter		N cetyl-trimethyl ammonium bromide (CTAB) (100 g)
Horizontal agarose gel system	Pipettes (5 ml, 1000 µl, 200 µl, 20 µl)		
Hotplate/stirrer			Orthoboric acid (boric acid) (500 g)
Digital camera analysis software (e.g. Kodak digital science 1D LE3.0)	Spatula		Poly vinyl pyrrolidone (PVP) (500 g)
	Stainless steel racks for 50 ml tubes		Silica gel, 35–70 mesh (500 g)
Shaker	Stirrer bars		Silica gel, self-indicating, coarse (500 g)
Ultracentrifuge (rotor specification: vertical 16 place, 6 ml polyallomer thin-walled tubes; max speed 65,000 rpm)	Thermometer		Sodium chloride (500 g)
	Transfer pipettes (3 ml)		Sodium hydrogen carbonate (500 g)
	UV transilluminator (312 nm)		Sucrose (500 g)
Water bath	Wash bottles		Tris (hydroxymethyl) aminomethane hydrochloride (Tris HCl) (500 g)

*values in brackets indicate the approximate quantity of each chemical needed to set up a DNA banking facility

Figure 12. An ultracentrifuge may be used for DNA extractions (photo: Molecular Systematics Section, Royal Botanic Gardens, Kew).

leaving the nucleic acids as a loose pellet at the bottom of the tube. The pellet is then 'washed' with ethanol, allowed to dry and re-suspended. The purpose for which the DNA is to be used can determine the method of re-suspension; for most molecular techniques, re-suspension in either deionized water or Tris-EDTA buffer is adequate. For DNA bank storage, however, it is better to undergo a more extensive process that provides further purification and allows for long-term storage of the DNA.

This process involves the re-suspension of DNA in an ethidium bromide/caesium chloride (EtBr/CsCl) solution and subsequent density gradient and dialysis purification. A concentrated solution (1.55 g/ml) of caesium chloride forms a density gradient after a few hours of high-speed centrifugation, where compounds are separated according to their density; DNA is concentrated as a distinct band within the gradient.

Ethidium bromide is used for direct visualization of DNA. The dye intercalates between the stacked bases of nucleic acids and fluoresces red-orange when illuminated with UV light, allowing the band containing DNA to be removed from the solution with a pipette. Ethidium bromide is then removed by adding butanol, in which the EtBr is more miscible than in water. The DNA then undergoes dialysis which removes the caesium chloride from the sample and stabilizes the DNA in a buffer which confines degradation.

The process of extraction and purification typically produces a highly purified DNA sample, which can then be stored at -80°C for an indefinite time. Assessment of the quality and quantity of the DNA extract can be made by either visualization on an agarose gel or spectrophotometry.

8. DNA banking and health, safety and security
E. Kapinos and S. Graham

Health, Safety and Security (HSS) considerations for DNA extractions and purification processes include good laboratory practice (e.g. wearing a lab coat and rubber gloves) and the provision of a safe working environment, with correctly functioning electrical equipment, non-slip floors, adequate lighting and air circulation. In addition, special precautionary measures must be taken when working with certain chemicals and specialized machinery and equipment (see examples in Table 4). Guidelines for the manual handling of heavy objects are also necessary. It is also important to note that some processes may not produce immediate harmful effects to the body when carried out once or twice but may lead to health problems when carried out on a regular basis over a longer period of time.

Where health risks can be anticipated, it is in the interest of both the employer and the employee to reduce the risk by promoting awareness and applying suitable control measures. Many countries will have specific legislation governing HSS issues at work. In Britain, for example, the Management of Health and Safety at Work Regulations (1992) place a general duty on employers to assess and control the overall risks to HSS posed by their activities. Health risks from the use of chemicals are part of the British regulations framework but are separately covered by the Control of Substances Hazardous to Health Regulations (COSHH; 1988, updated 1994 and consolidated in 2002). These regulations aim to protect health by preventing or controlling exposures to chemicals and other substances. They also demand that a risk assessment is carried out for each work-related activity that takes place in a laboratory, serving as a baseline for any control or protection measures to be undertaken. The initial risk assessment should also include risks of potential exposure to other hazards (e.g. flammable substances, vacuum), physical agents (e.g. noise, ionizing radiation) and ergonomic stresses (such as those posed by manual handling of loads or the use of display screen equipment), as outlined in The Royal Society of Chemistry's *COSHH in Laboratories*.

8.1. Personal protection equipment

Engineering controls (such as the use of ducted fumehoods) will not prevent splashes or powdery chemical substances coming into contact with skin or eyes. Therefore personal protection equipment (PPE) is essential for any work in the laboratory. A lab coat with a high collar and elasticised cuffs not only protects personal clothing but also shields exposed skin from potential hazards, such as ultraviolet (UV) radiation when removing DNA bands after a density gradient run. For that task, a visor covering the whole face and throat should be worn as well. Recommendations for lab coats also include a breast pocket deep enough to hold protective glasses, a flap at the rear for free leg movement and knee-covering length when the wearer is in sitting position. Traditional cotton twill lab coats provide sufficient protection for most laboratory tasks. Closed shoes will also give protection to feet in case of any spillages and are required in a laboratory environment.

Inhalation of fumes or dust from hazardous chemicals poses the most frequently encountered risk to health during DNA extractions and related tasks. All work that harbours such risks should be carried out in a ducted fumehood. Only in situations where this risk is small are face-masks covering the mouth and nose considered adequate protection, for example when weighing out plant material in silica gel.

Gloves are essential for any work with hazardous materials and are recommended in most cases when handling chemical or powdery substances to protect against harm to the skin, absorption of any substance and allergic responses. For protection against dry solid materials general purpose lightweight laboratory nitrile and vinyl gloves are sufficient. For work with liquid chemicals, however, they only provide limited protection as breakthrough times for different chemicals vary. Therefore, gloves are considered to offer only splash protection when in contact with liquid chemicals and should be cleaned or replaced immediately after chemical contamination. For the handling of some toxic materials (e.g. ethidium bromide), wearing a double layer of gloves and exchanging them after about 20 minutes is recommended to give additional protection. Gloves are highly recommended for any work with equipment emitting UV light, as they give good protection against the full range of wavelengths. For full hand and forearm protection, the lab coat sleeve should be tucked into the gloves during UV light exposure. For work with ultra low temperature freezer samples and liquid nitrogen operations, thermal gloves should be worn. Gloves can themselves cause allergies, especially powdered gloves. In such cases either a pair of cotton gloves can be worn underneath the latex gloves, or blue/purple non-powder, non-latex gloves may provide a solution.

Eyes need to be protected against dust and possible splashes from harmful chemicals and other substances (such as blood). Although the risk of splashes is small when working with minute quantities of a solution, it still

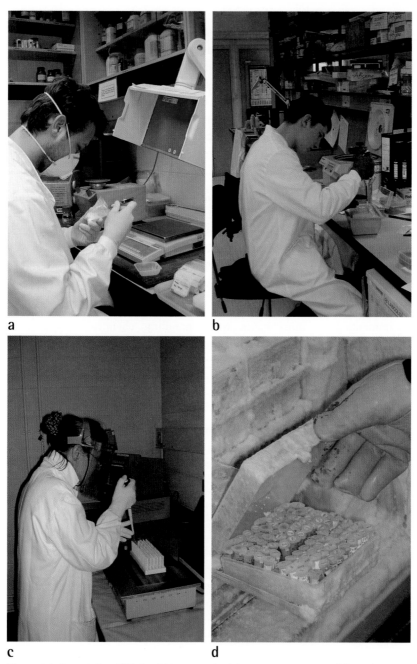

Figure 13. Appropriate PPE should be worn for laboratory procedures: (a) face mask when handling material with silica gel powder; (b) gloves and lab coat for routine work; (c) full face shield when examining DNA under UV light; (d) insulated gloves when handling samples stored in ultra-cold conditions (photos: Molecular Systematics Section, Royal Botanic Gardens, Kew).

exists, so eyes should be shielded, especially when dealing with toxic or corrosive chemicals. Safety spectacles, or better still a full-face visor, are appropriate when performing fiddly tasks, such as removal of tubes filled with ethidium bromide from an ultracentrifuge. Emergency eyewash equipment must be provided in every laboratory.

When working with potentially contagious material, such as tissue samples or blood, laboratory technicians should use biosafety cabinets, with a laminar airflow and UV sterilization.

8.2. Hazards arising from DNA extractions and related processes, especially use of ethidium bromide

DNA extractions present potential HSS hazards at various stages: (i) preparing the plant material (either fresh, dried in silica gel or from herbarium specimens) for extraction; (ii) preparing chemicals required for extraction and purification process; (iii) carrying out the extraction process, for example for density gradients and dialysis; and (iv) carrying out electrophoresis. Table 4 gives an overview of the potential hazards, typical route of exposure, degree of risk and possible control measures.

A number of different chemical substances are required for DNA extraction, density gradients, dialysis and agarose gel electrophoresis. These substances include mixtures and preparations as well as pure substances, all with different activities. When assessing hazard categories for a multiple substance activity, preference must be given to the most hazardous substance. Hazard categories range from low, moderate and high (very toxic) to special category (which includes extreme hazard and substances with carcinogenic or mutagenic activity or with toxicity to reproduction). In Britain, these hazard categories are part of the classification scheme within the Chemical (Hazard Information and Packaging for Supply) Regulations; classifications are included on bottle labels and suppliers' product safety datasheets and will therefore be the first source of information available to users.

Density gradient ultracentrifugation of DNA involves the preparation and use of an ethidium bromide/caesium chloride (EtBr/CsCl) solution. This solution is used for the re-suspension of dried DNA pellets (which must be carried out in a ducted fumehood) as well as for the ultracentrifugation run. Ethidium bromide is a fluorescent (orange/red) dye, which intercalates into the DNA structure. As this may cause heritable genetic damage, ethidium bromide on its own, especially in powder form, is considered to be a high-risk chemical substance. It is therefore advisable to allocate a specifically designated working space. It is essential to confine all ethidium bromide related tasks to a separate room that can also accommodate the ultracentrifuge. The latter requires a protective, enclosed space around it due to its high-speed rotations (up to 60,000 rpm), as well as its contact with EtBr/CsCl during the gradient runs. Any tools and equipment should be confined to this room. Contamination of work surfaces can be kept to an acceptably low level through monthly cleaning with 70% ethanol and covering with strong tissue paper (changed regularly). Double gloves and lab coat must always be worn when directly handling the solution, and a full-face visor, or at least eye protection, is recommended for certain tasks such as visualization of DNA under UV light or removing DNA from gradients.

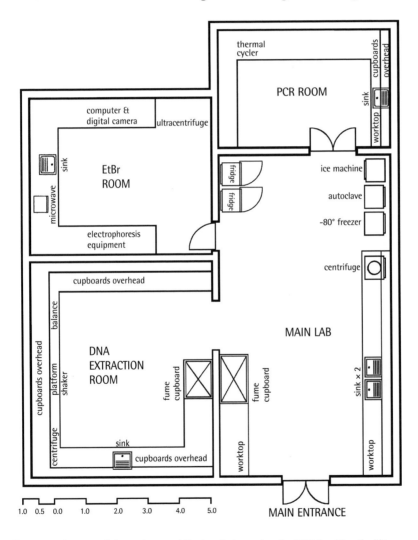

Figure 14. Layout of the various architectural elements of a DNA banking facility.

Box 8 Laboratory facilities F. Conrad and K. Balele

To avoid contamination, it is preferable that DNA extractions take place in a separate room from the main laboratory where other DNA work is carried out and also away from the DNA bank storage room.

For HSS reasons the laboratory should have a ducted fumehood, to be used when measuring out certain harmful chemicals involved in plant collection and DNA extraction, particularly silica gel, detergents (e.g. CTAB) and ethidium bromide when in powdered form.

A separate room, the 'gel room', is also essential for procedures involving the use of ethidium bromide, to avoid contamination with the other parts of the laboratory. These include the use of agarose gels stained with ethidium bromide and associated electrophoresis equipment, preparation of EtBr/CsCl density gradients and use of the ultracentrifuge. Samples, once they have been subject to ultracentrifugation, also need to be processed in this room, both to visualize the DNA bands under a UV light and to prepare the samples for dialysis.

A double-distilled water purification system is essential for preparation of solutions, in addition to an autoclave for the sterilization of double-distilled water, pipette tips, microcentrifuge tubes and some solutions and buffers.

In order to purify DNA extracts for long-term storage an ultracentrifuge and rotor, capable of running at 54,000 rpm, is usually required. This represents by far the most costly investment in the establishment of a bank of extracted DNA.

Perhaps the second largest piece of equipment, in terms of size and cost, is a -80°C freezer for long-term storage of the DNA extracts. Other smaller but essential items of equipment and consumables required to follow the 2× CTAB DNA extraction protocol (Appendix 1) are listed in Table 3. The total equipment cost is about US$90K or €70K.

A DNA bank needs at least one full-time DNA bank manager. The collection of material from the wild, and subsequent DNA extraction and storage, involve a series of tasks that cumulatively contribute to the quality of DNA archived for future research. Proficiencies in the collation and databasing of collection information, preparation of vouchers and general laboratory procedures are all necessary to maintain a high standard of quality control. The DNA bank manager also acts as a gate-keeper to prevent any unauthorised access to the storage area and thus prevent any misplacement, accidental thawing or contamination of the samples.

Table 4 Health, safety and security measures associated with activities carried out in a DNA bank

Activity	Hazard	Route of exposure	Risk	Control measure
Silica gel handling				
Mixing fine mesh and indicating silica gel; measuring silica into ziplock bags for sample collection; removing plant material from silica gel	Silica gel, 22Å, (28–200 mesh) Silica gel, type III, colour-indicating granules Dust from dry plant material	Inhalation Contact with eyes	MOD	Wear gloves and eye protection Work in a ducted fumehood or wear a face mask Always use a ducted fume hood when removing samples from containers, and when working with plant material known to be toxic, or where toxicity is unknown
DNA extraction				
Preparing and using extraction buffers	CTAB Tris HCl EDTA NaCl ß-mercaptoethanol PVP	Ingestion Inhalation Contact with skin or eyes	MOD	Wear suitable PPE Work in a ducted fumehood in designated room Wear a face mask when weighing powders Dispose of contaminated pipette tips in a waste container stored in a ducted fumehood
Grinding plant material with silica gel and CTAB	Silica gel, 22Å, (28–200 mesh) CTAB	Contact with eyes Ingestion Contact with skin	LOW	Wear suitable PPE Work in a ducted fumehood in designated room Wear a face mask when weighing out dried plant material
Using SEVAG to emulsify plant tissue	Chloroform Isoamyl alcohol	Ingestion Inhalation Contact with skin *Prolonged exposure has risk of irreversible effects and serious damage to health* Inhalation *Flammable*	MOD	Wear PPE Work in a ducted fumehood in designated room
Precipitating DNA and spinning down precipitate	Ethanol Isopropanol	Ingestion Contact with skin *Highly flammable*	LOW	Wear suitable PPE Work in a ducted fumehood Keep away from all sources of ignition and static charges
Density gradient ultracentrifugation				
Preparing EtBr/CsCl solution; resuspension of DNA pellet; ultracentrifugation; removal of tubes from ultracentrifuge rotor and DNA bands from tubes continues	EtBr	Ingestion Inhalation Contact with skin or eyes *Very toxic by inhalation and possible risk of irreversible effects* *May cause heritable genetic damage*	HIGH	Wear suitable PPE, including a double layer of gloves Work in designated room, and dispose of all waste in that room Work in a ducted fumehood when weighing out EtBr and when resuspending DNA

Table 4 continued				
Activity	Hazard	Route of exposure	Risk	Control measure
	CsCl Tris HCl EDTA	Ingestion Inhalation Contact with skin or eyes		Wear a full face visor when removing tubes from the ultracentrifuge and when removing DNA bands under UV light

Dialysis

Preparing samples for dialysis	Butanol	Inhalation *Highly flammable*	HIGH	Wear suitable PPE, including a double layer of gloves Work in a ducted fumehood Dispose of all waste in designated room
	EtBr	Ingestion Inhalation Contact with skin or eyes *Very toxic by inhalation and possible risk of irreversible effects* *May cause heritable genetic damage*		
	CsCl Tris HCl EDTA	Ingestion Inhalation Contact with skin or eyes		

Agarose gel electrophoresis

Preparing agarose gel; electrophoresis of DNA samples	EtBr	Ingestion Inhalation Contact with skin or eyes *Very toxic by inhalation and possible risk of irreversible effects* *May cause heritable genetic damage*	HIGH	Wear suitable PPE Work in designated room where possible Wear a full-face visor or safety goggles when visualizing DNA under UV light
	Agarose	Inhalation Contact with eyes		
	Boric acid	Inhalation Contact with skin or eyes May cause irreversible effects		
	Tris HCl EDTA	Ingestion Inhalation Contact with skin or eyes		

Box 9 Shipping specimens and tissues A. Corthals

Specimen and tissue transport across borders may require export and/or import permits and phytosanitary certificates. Regulations vary from country to country and according to the type of material. The shipping method used may also introduce another level of permitting, depending on whether the specimens have been shipped fresh, frozen, or preserved in fluid. A copy of the applicable permits and/or phytosanitary certificate containing the identification and description of the purpose of the tissue should be attached inside the package and outside with the shipping label. Herbarium specimens do not generally need phytosanitary certificates (it is wise to check) but may require export permits and/or CITES permits.

Vertebrate tissue samples (e.g. blood and muscle) are routinely transported in fluid (ethanol) or frozen (in dry ice or liquid nitrogen). Transport of frozen tissue on dry ice or in liquid nitrogen is simple when air shipment is not required. However, authorisation for shipment of these chemicals on airlines is mandatory because both are classified as Restricted Articles. Policies for carrying such materials vary with the carrier; check beforehand. Special black-and-white stickers must be added to the container for acceptance as air cargo. Airlines do not allow nitrogen to be transported in a liquid state, but allow its transport in a non-pressurised 'dryshipper' Dewar container, which absorbs liquid nitrogen and releases the vapour only, hence its name 'dry'. The package must be clearly labelled 'non-pressurised dryshipper', 'this side up' and 'do not drop: handle with care' to prevent careless handling. The dryshipper must be accompanied by a letter of exception (usually delivered by the manufacturer) explaining that the Dewar does not contain nitrogen in a liquid state.

Some samples can be shipped dried (such as plants, skin, hair and blood on paper), with a small amount of desiccant (e.g. silica gel) and placed in airtight containers to prevent rehydration.

The most convenient method for transporting tissue samples of most invertebrates, vertebrates and marine algae, is preservation in 70–100% ethanol. However, since 9/11, carriers have new and tighter restrictions on the amount of ethanol allowed in the cargo area. You are advised to check with the airline about the amount of ethanol allowed on board and in cargo. Ethanol is flammable and highly volatile and should therefore be stored in unbreakable, leak-proof plastic vials. Sealing the vials inside thick plastic sleeves is advised as a further precaution.

9. Databasing and information technology
S. Cable and T. K. Fulcher

Before starting a DNA bank database, it is important to decide whether to allocate major resources and staff time to database development, or to DNA banking and research. This may be less of an issue for institutions with a large budget and the support of a strong information technology (IT) department, but smaller projects could find developing a database a significant drain on resources. Other considerations include the expertise of staff, the anticipated size of the database in terms of specimen records, whether the database will be run on a network of computers, the number of concurrent users, levels of access and security and the types of outputs such as queries, text reports or web interface. Before choosing database software, shop around and if possible try some systems and talk to users as well as developers.

9.1. Scope

Depending on the requirements of the project, a DNA bank database could comprise anything from a single table recording DNA transactions to a comprehensive relational database incorporating taxonomic and collection data. A simple database will list DNA samples (and information such as accession number, date of extraction, quantity, quality and sequence data deposited in a central repository, e.g. GenBank), transfers (date, quantity and recipient) and any restrictions that have been placed on the use or transfer of the material (from collecting permits, MTAs or other agreements). These transaction data could be part of a larger taxonomic or collections orientated database, with interlinked information on taxonomy (such as taxon name, synonymy, references, habit, ecology and distribution) and collection data for the supporting voucher specimen or e-voucher (such as collector's name and number, image name, grid reference, locality, description, collection date, identifying taxonomist and identification date). As an example, the schema of the tissue bank database at the American Museum of Natural History is illustrated in Figure 15.

9.2. Software

There are several options for starting a DNA bank database, including adapting an existing system (if your institution has one), using an off-the-shelf system or developing your own. Arguably the most practical and affordable option for smaller projects is to use a ready-made taxonomic and collections management database system. Some starting points for looking for these are the web pages of Digital Taxonomy and the Internet Directory for Botany (see www.geocities.com/RainForest/Vines/8695/software.html; www.botany.net/IDB). Many botanists use the BRAHMS database (see http://herbaria.plants.ox.ac.uk/bol/home/). BRAHMS (Botanical Research And Herbarium Management System) includes a module for tracking DNA samples, sequences and source material. It facilitates export of sequence data to analysis programmes and can provide links to GenBank through the web page using accession numbers. However, it does require a Microsoft Windows environment and a number of Windows-specific library files that can be downloaded from the BRAHMS website.

There are two main options for developing your own DNA bank database program from scratch: either (i) use commercial software, for example Access, Foxpro, Sybase or Oracle, or (ii) use free or open-source software under one of the public licensing schemes, such as the GPL (General Public License), for example MySQL or PostgreSQL. Microsoft Access is a good solution for small single-user databases due to its ubiquity and ease of use. However, developing Access beyond a few interlinked tables requires programming skills, and for databases with large numbers of records and multiple users on networks there is other database development software that offers higher performance, such as Foxpro. Alternatives such as Oracle and Sybase are examples of high-end development packages for corporate databases on large networks. They are expensive in terms of licenses, support and the programming expertise needed to develop applications. Open source software, such as MySQL, offers a cheaper alternative, and although the services of programmers may be required, there are online communities of programmers that freely exchange ideas and solutions to problems (see www.mysql.com for an example).

9.3. Design

Whichever route is chosen, the actual data will be contained in a series of individual tables within a relational database (see Figure 15). A table is simply a set of rows and columns; the rows represent records (e.g. the taxon, collection or DNA sample) and the columns are the fields of data (e.g. family, genus, species, collector's name, collector's number, accession number, country, gazetteer or locality). Microsoft Access includes wizards to simplify the process of creating and linking tables from scratch or to create a relational structure from single tables of existing data. Tables are linked through the use of unique identifiers (normally numbers) in specific fields that are not available to users for editing. The objective is the elimination of redundant data and the logical organisation of related data within tables for efficient data storage, easy updating and optimal performance.

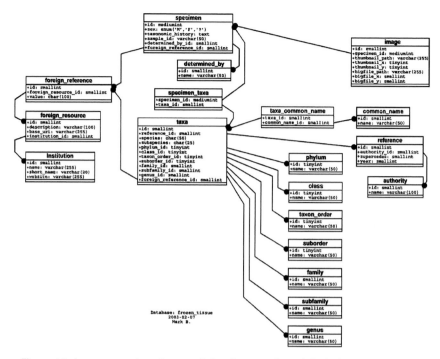

Figure 15. A representation of some of the elements that might be incorporated into a DNA banking database system (schema of AM-CC database by Mark Breedlove).

There are a few basic design principles:

(i)   create unique atoms of data (i.e. there should only be one version of a collector's name or a locality)

(ii) remove fields that duplicate data within a table

(iii) create separate tables for subsets of data that apply to multiple records (e.g. separate tables for collectors' names and species names etc.)

(iv) logically link related tables (e.g. the table of authors' names should link to the table of species names and not the table of collection data)

(v)   ensure that multiple instances of an atom of data (e.g. a collector's name) in a parent table (e.g. the collection data table) refer to single instances in child tables (e.g. the collector's name table).

The most user-friendly databases hide much of the complexity from the user and present a familiar Windows-feel to the interface. Tables usually provide a quicker mechanism than forms for entering or editing records.

Think of the poor user when your database design requires them to manually enter accession numbers to access each record and your beautiful cascading forms have to be opened and closed repeatedly. Any redundant keyboard strokes and mouse clicks soon add up to a considerable waste of staff time if your database grows to many thousands of records. Each atom of data, such as a species name, should only ever have to be typed once into a well-designed database. All subsequent instances of the same data should be selectable from pick-lists. Standard lists of families, genera, authors and localities are available online and could be incorporated into your database as pick-lists. Ideally, a database may be configured to reduce typing when entering new records by automatically incrementing the accession or collection number and copying repeated information, such as the taxon name or locality data, from the previous record (this is an option in BRAHMS, for example).

In large systems with multiple tables, direct access to the data is not an option, as editing records would require detailed knowledge of the relationships between tables. The database application should only allow access to the necessary fields, and when the 'update' button is selected, the appropriate data should automatically be updated throughout all of the tables. Heavy concurrent use may require a mechanism for locking records to avoid redundant edits and loss of data resulting from simultaneous updates by different users. Some large institutions will require a database solution based on the Unix operating system. The interface for the user could be on a Windows desktop, but the database may sit on a server somewhere else in the building or perhaps off-site. Individual researchers and projects should take expert advice on concurrent multi-user databases on Windows and Unix networks.

9.4. Security

Security is a major consideration for any database beyond the scope of a single researcher on an individual computer. The issues include protection of intellectual property and data integrity as well as observing obligations under the CBD, collecting permits or agreements. Institutional networks provide a minimum level of security, but databases should have built-in levels of access that database managers can assign to users. Access to the database, whether read-only or for editing, should be constrainable to particular subsets of the records or to particular fields within the records. Access for collaborators can be provided over the internet and should be securely password-protected.

9.5. Data standards

Data standards are a critical component of any database project, but their importance is often underestimated. The project data standard defines the fields used (name, size and type), the information stored in each field (e.g. collector's name, species epithet or grid reference) and the format of that information (e.g. whether a collector's name is written as 'Smith, D. G.' or 'Smith D.G.' or 'D.G.Smith'). It may also indicate the syntax and punctuation to be used within fields (e.g. whether to capitalize the first letter and end with a full-stop).

Consistent data standards underpin the value of a database as a research tool and the exchange of data between databases, research partners in country of origin and projects. Basic queries, such as how many taxa are stored in a DNA bank, can only be answered if each atom of data, such as a taxon name, is unique with no non-standard variations. Think of a database as a publication-quality report generator, except that instead of the information stored as one long fixed text file, it is broken down into small reusable chunks that can be reordered and reassembled into a variety of outputs such as voucher labels, transaction lists, inventories, maps or even monographs.

10. Collection of plant material for DNA extraction and voucher specimens
A. Paton

10.1. Collection and drying techniques

Traditionally, plants have been preserved for scientific research by pressing and drying. This is done by pressing the sample between sheets of absorbent drying paper in a plant press. The material will dry more quickly if the press is placed over a heat source and the drying paper is changed regularly; herbarium specimens that are dried quickly are suitable for DNA extraction. Alternatively, specimens are placed between sheets of paper, compressed and tied. The bundle is then placed in a heavy gauge polythene bag with sufficient 70% alcohol to soak the specimens; the material can then be pressed and dried at a later date, but this type of preparation is far from ideal for subsequent DNA extraction. Pressed and dried specimens, derived from either method, are then stored in a herbarium. A good account of the equipment needed and the processes involved in pressing a plant so that it can serve as a herbarium specimen can be found in Bridson and Forman's *Herbarium Handbook.*

Although these methods are effective for the preparation of herbarium voucher specimens and DNA has been effectively extracted from rapidly dried herbarium specimens, the drying time involved is usually well over 12 hours, and some degradation of DNA is likely. Therefore, material from which DNA is to be extracted is best collected into silica gel while the tissue is fresh. In this method, 1–2 grams of fresh leaves are torn into small pieces and placed in a small ziplock plastic bag along with at least 10 times the weight of silica gel. The aim is to dry the leaves quickly, in under 12 hours, to preserve the DNA. Fine mesh silica gel is more effective at drying leaf tissue than coarser sizes. Orange indicator crystals that turn pale when the silica gel is saturated should be added. This helps ensure that the silica gel used will be an effective drying agent. After the leaf material has been dried, most of the silica gel can be removed and if not yet saturated with water it can be reused. Drying the silica gel in an oven (175°C for one hour) will regenerate it, allowing reuse. If silica gel is to be reused, care must be taken to ensure that no fragments of previously dried tissue contaminate the silica gel. Note that in instances for which DNA extraction proves problematic, perhaps due to slow drying of succulent leaves or due to some aspect of plant chemistry, extraction from floral or other tissue may be more effective than leaf material.

10.2. Voucher specimens

A voucher can be defined as a preserved specimen and its associated information that physically documents the existence of that organism at a given place and time. Within the context of DNA banking, the voucher specimen is evidence of the organism from which the tissue was originally removed. Without voucher specimens, there is no way of knowing whether the sampled organism has been correctly named, suspect results cannot be checked and the experimental work lacks scientific rigour as results cannot be unambiguously duplicated: therefore, proper vouchers for every DNA sample are essential for a DNA bank.

If a voucher is to serve as evidence for the identification of a sample, it must carry sufficient information. Thus specimens should be fertile, if possible displaying flowers and fruit. Underground parts should be included as part of the specimen, or details of this noted on the accompanying label. Sufficient leaf and stem material should be collected to display features of these organs. Characters that are lost after pressing should be captured, for example by providing a description on the specimen label or creating an e-voucher. Such characters include habit (herb, shrub or tree, height and general appearance), flower colour, fruit colour, smell of plant and presence and colour of sap. Occasionally a researcher may wish to extract DNA from a sterile plant. A voucher specimen should still be prepared from the plant from which DNA is to be extracted, although it may be difficult to verify the identification of the species.

The label information should also include details of collection locality, with sufficient detail to allow recollection, date of collection, habitat notes, GPS positioning and elevation. This information can be useful in identification and can also help understand genetic variation within a species. If the sample is being taken from a cultivated plant, then the cultivated collection catalogue or accession number of the plant should be recorded. However, it is insufficient only to record this number without also taking a voucher specimen and recording the information attached to the plant, such as collection details. Plants in gardens and other collections can be misnamed, die or be moved and also catalogues may not be maintained. A specimen is required to provide a permanent record.

As much as possible, the voucher specimen should be taken from the plant at the same time as the tissue sample. This ensures that the voucher accurately reflects the sampled plant. It is important that the DNA sample in the DNA bank and the voucher specimen are cross-referenced and the DNA bank accession number should be recorded on the voucher specimen label.

If collecting material in the field, take care to ensure that the material is collected legally. Permits should cover the collection of herbarium voucher specimens as well as the DNA sample. The herbarium voucher specimens, if incorporated into a national herbarium, may be consulted by many researchers over time, sent on loan to others, further sampled for DNA or phytochemicals and perhaps imaged and databased, and this information should be made publicly available. The collector needs to ensure that the prior informed consent (PIC) covers collection and future use of vouchers. It may be necessary to collect duplicate vouchers because the country of origin may request that one set of specimens be kept in that country. If the country of origin places restrictions on the use of the voucher material, such as saying the vouchers may not be sent on loan to another institution, then these restrictions should be noted on the specimen itself and in the management system of the herbarium housing the specimen.

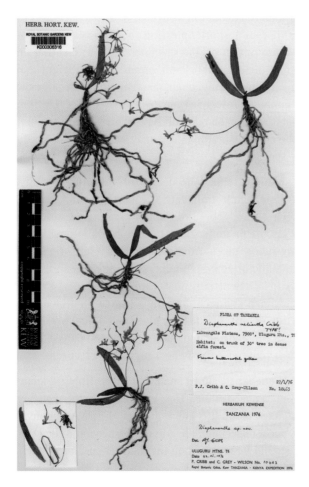

Figure 16. Plant herbarium specimen (photo: API project, Herbarium, Royal Botanic Gardens, Kew).

10.3. Maintenance and accessibility

Voucher specimens should be stored permanently, and the specimens should be accessible to researchers who may want to study them. Specimens should be kept at a stable relative humidity of 40–60% and in a pest-free environment to prevent degradation. Ensuring specimens are housed and managed under suitable conditions and ensuring their accessibility is a considerable overhead cost. For these reasons, voucher specimens are often deposited in already existing herbaria. If DNA bank vouchers are to be housed by a herbarium under different management from that of the DNA bank itself, then an arrangement to deposit and house the vouchers should be agreed prior to sending the specimens. The cross-reference between DNA bank sample and voucher specimen should be preserved. If the annotation of a voucher is altered, for example due to a misidentification, then the change in name should be communicated to the DNA bank. The easiest way to ensure such changes are recorded is to database the voucher specimens. Such a system would also allow recording of any restrictions placed on the use of the voucher. This recording of information is another cost that needs to be considered. Collaboration between the herbarium holding the voucher specimens and the DNA bank is important in ensuring that voucher specimens can serve their function.

Box 10 **Animal voucher specimens and e-vouchers** A. Corthals

To quote from a review by Prendini and colleagues (2002), the nature and acquisition of animal voucher specimens is dependent on the taxon in question: 'For many taxa, diagnostic features of the specimen used as a tissue sample may suffice and may be all that is available if only a single specimen could be obtained for molecular analysis. However, it is preferable to retain a second, intact specimen as the voucher ... Additional specimens, representing sexual and ontogenetic variation, should always be acquired, if possible.'

Just as with tissue and DNA samples, long-term storage of specimens is of prime importance and requires techniques of best practice in conservation. Without proper archival methods, collections run the risk that the identifying characteristics of improperly curated specimens may be damaged or lost, just as DNA might degrade.

Standards of curation such as storage, labelling and data recording should be applied and adapted to the needs of specific taxa. Thus, whereas skeletal vouchers must be stored dry, soft-bodied vouchers (such as worms) must be stored in ethanol or formaldehyde.

Box 10 continued

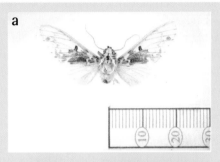

Figure 17. Examples of e-vouchers: (a) AMCC 112692, *Idalus intermedia*, Ecuador, collected by Suzanne Rab Green. E-voucher (photo) by Suzanne Rab Green; (b) Malagasy children holding the catch of the day (cat/leopard shark), which also happens to have been sampled by Phaedra Doukakis for tissue deposited at the American Museum of Natural History (AMCC 103658). This last image is a good illustration of the e-voucher concept. While it is clear that the picture cannot reveal all morphological characteristics, it is the only visual record remaining of the whole specimen, as the fish was to be sold and consumed (photo: Phaedra Doukakis, Wildlife Conservation Society/Pew Institute for Ocean Science).

The term 'e-voucher' has been defined by Monk and Baker (2001) as follows: 'An e-voucher is a digital representation of a specimen. An e-voucher may be ancillary to a classical voucher specimen or it may be the only representative of the specimen in the collection.' Traditionally, voucher specimens deposited into museum collections were morphological vouchers, and in most cases consisted of the cadaver of the whole animal. In the case of a frozen tissue repository, this traditional definition of a voucher becomes not only impractical, but for the majority of animals sampled, simply impossible to apply. Thus, the digital picture of the specimen sampled becomes the morphological voucher in the many cases where the remains of the animal sampled are unobtainable. To that effect, whenever possible, the specimen records should be linked to digital images, making for a complete connection between tissue samples and subsequent DNA sequence data and the visual identity of the specimen examined.

11. Genetic resources, systematics and the CBD: a case for DNA banking in South Africa

G. Reeves, J. C. Manning, P. Goldblatt, D. Raimondo and M. Wolfson

This chapter outlines the efforts and underlying motivation of the South African National Biodiversity Institute (SANBI) to establish a DNA bank of the South African flora (which comprises approximately 22,000 species in 2,000 genera). Funded by the Darwin Initiative of the UK's Department for the Environment, Food and Rural Affairs (www.darwin.gov.uk), and in collaboration with the Royal Botanic Gardens, Kew, this initiative was conceived to assist South Africa to meet its obligations under the Convention on Biological Diversity, thus recognizing in the genetic era the relevance of DNA banking and systematics to the CBD.

The year 2003 signalled fifty years since the discovery of the double helix, and during this time almost no aspect of biological science has remained untouched by the revolution in molecular biology. Taxonomy has been no exception, and the generation of DNA sequence data for phylogenetic reconstruction is now commonplace. However, despite the widespread integration of molecular systematics into taxonomy, along with the acceptance that taxonomy underpins successful implementation of the CBD, many if not most current museum and botanic garden taxonomic collections are inadequate for long-term, high-quality DNA extraction and preservation. The South African in-country DNA banking initiative now allows plant DNA to be extracted and stored long-term and transferred within the global academic community according to provisions set out in a Material Transfer Agreement (MTA). We believe that this endeavour is especially relevant in the new biotechnological age and may stand as a case study for other biodiversity-rich developing countries as a means to regulate and monitor access to DNA for the benefit of the global academic community – in a way that satisfies the CBD whilst still encouraging academic exchange.

11.1. An introduction to plant species diversity in South Africa

The distribution of plant species across sub-Saharan Africa is highly uneven, with regions of high concentration of species contrasting with a relative paucity in others. At an African level the flora of southern Africa is the most diverse – with some 22,000 species of seed plants, it is far richer than the floras of either East Africa, with fewer than 12,000 species, or West Africa, with around 7,000 species. The flora of the whole of

Figure 18. The Kirstenbosch Research Centre, surrounded by Fynbos vegetation at the foot of Table Mountain in Cape Town, hosts the DNA banking facility of SANBI (photo: SANBI).

tropical Africa, encompassing an area of 2.5 million km², is estimated to be between 30,000–40,000 species. Southern Africa, covering an area of 2.5 million km², thus accounts for roughly 50% of the sub-Saharan flora in just 10% of the land area.

In general terms, the level of biological diversity decreases from the equator towards the poles. In Africa, this decrease is less marked in the southern hemisphere than in the northern. Plant diversity at the Tropic of Capricorn is around half that at the equator and drops to one-third by 30°S. By 34°S, however, the level rises again to around that recorded at 20°S. This reversal in the global trend is the anomaly that constitutes the Cape Floristic Region (CFR) in extreme southwestern Africa. Although only around 50–75% as rich as the richest tropical African floras (measured as species recorded per 2.5 degree grid), the Cape far surpasses them in its level of endemism.

One of the characteristics of the Cape flora is its unusual family composition. The two largest families, Asteraceae (daisies) and Fabaceae (legumes), typically dominate the floras of semi-arid regions, and together comprise some 20% of the total species in the Cape. Unique to the region, however (and consequently to the southern African flora as a whole), is the significant contribution made by Aizoaceae, Ericaceae and Iridaceae.

Figure 19. Cape flora (photo: John Manning).

Scrophulariaceae, Proteaceae and Restionaceae follow in size. The importance of Ericaceae, Proteaceae and Restionaceae in the Cape flora, both in terms of biomass and species diversity, is widely appreciated, but the remarkable diversity of Iridaceae, predominantly a family of herbaceous, seasonal geophytes, is especially striking. Nowhere else in the world does this family provide more than a small proportion of the total species. The adaptive radiations of Ericaceae and Iridaceae in the Cape flora are one of its most unusual aspects.

The unusually high species richness of the Cape flora is generally interpreted to be the result of more or less uninterrupted evolution of the flora in a region of high physical complexity, the consequence of sustained post-Pliocene climatic stability. Molecular phylogenetic trees provide a powerful tool for assessing the history of this unique flora. The integration of patterns of diversification with palaeoclimatic data is likely to provide significant insights into the forces that have influenced the current characteristics of the flora. At present, for example, it is unclear whether the remarkable pattern of radiation without generic diversification in Cape Ericaceae is associated with the relatively recent arrival of ancestral stock in the Cape or with rapid evolution following the establishment of a Mediterranean climate at the Cape. The answers to this and similar questions are now at our disposal via the use of molecular techniques and modern methods of phylogenetic reconstruction. The South African DNA bank will serve as a vital resource in these endeavours, allowing the effort that is invested in plant collection to be maximised by making DNA extracts available for use by the academic community at large.

11.2. Bioprospecting, indigenous plant use and sustainable livelihoods in South Africa

In South Africa, indigenous plant species contribute significantly to people's livelihoods and are extensively used for fuel, building materials, food and medicines. The use of plant species in South Africa has never been fully quantified, but interesting case studies from the medicinal trade show that the number of species being utilized is probably substantial, with just over 700 plant species being actively traded for their medicinal properties. At present it is estimated that traditional medicines (mostly derived from indigenous plant species) are used by 27 million South Africans, which constitutes just under 80% of the population.

Commercialisation of indigenous plants for horticultural, medicinal and cosmetic trade, both nationally and internationally, is currently being explored as a means to alleviate poverty in rural areas and contribute to the socio-economic development of South Africa. Plants such as devil's claw (*Harpagophytum procumbens*, Pedaliaceae), used to treated rheumatism, have over the past three years provided a source of cash income to over 3,000 South Africans whose previous livelihoods depended solely on subsistence goat farming. Other indigenous plant products that are increasingly being traded on the international market include marula (*Sclerocarya birrea*, Anacardiaceae), used to produce marula beer, marula juice and marula oil for cosmetics and wild harvested indigenous teas including the legumes rooibos (*Aspalathus linearis*) and honey bush (*Cyclopia* spp.). Within the CFR the unusual floral diversity is being exploited for the cut-flower trade, with 82 species being harvested from the fynbos. Trade in indigenous cut flowers contributes to the livelihoods of both fynbos farmers and poorer communities, who harvest from both communal and private land.

Ensuring sustainable use of South African plant species, while at the same time contributing to the country's socio-economic development, is a massive challenge for the conservation sector of the country. Currently many popular medicinal plants traded on the local markets are suffering from over-exploitation, with many populations of species such as wild ginger (*Siphonochilus aethiopicus*, Zingiberaceae) and the pepper-bark tree (*Warburgia salutaris*, Canellaceae) becoming locally extinct. Determining species vulnerability to use, setting sustainable harvest methods and quotas and ensuring that value addition involves the equitable sharing of trade-derived benefits are all areas where significant attention is currently being directed, and all can be helped and monitored via DNA banking and molecular techniques. South Africa's Biodiversity Act includes a strong focus on access and benefit-sharing in addition to appropriate commercialisation and sustainable use of biodiversity. Perhaps most notably, as part of South Africa's commitment to the CBD, the National Biodiversity Strategy and Action Plan is currently developing the means to ensure that this area receives sufficient resources and attention in the future.

11.3. South African legislative framework

At the end of May 2004, the Biodiversity Act No. 10 of 2004 (Government Gazette Vol. 467, 7 June, 2004) was signed by the President. The Act provides for the management and conservation of biodiversity, the use of indigenous resources in a sustainable way and the fair and equitable sharing among stakeholders of the benefits arising from bioprospecting of indigenous biological resources. The Act also gives effect to international agreements relating to biodiversity that South Africa has ratified, provides for co-operative governance in biodiversity management and conservation and establishes the South African National Biodiversity Institute to assist in achieving the objectives of the Act. In terms of the Act, 'biodiversity' refers to the variability among living organisms from all sources and includes diversity within species, between species and of ecosystems, whereas 'indigenous biological resources' refers to any living or dead animal, plant or other organism of any indigenous species, any derivative of, or any genetic material of such animal, plant or other organism.

The State is regarded as the trustee of biological diversity, and as such has the responsibility to manage, conserve and sustain South Africa's biodiversity and its components and genetic resources. Chapter 6 of the Act, entitled 'Bioprospecting, Access and Benefit-Sharing', sets out the framework for the regulation of access and benefit-sharing in South Africa. Its purpose is to: (i) regulate bioprospecting involving indigenous biological resources; (ii) regulate the export from the Republic of South Africa of indigenous biological resources for the purposes of bioprospecting or any other kind of research; and (iii) provide a fair and equitable sharing by stakeholders in benefits arising from bioprospecting involving indigenous biological resources.

The indigenous biological resources identified in this Chapter of the Act include any indigenous biological resources, whether gathered from the wild or accessed from any other source, including: any animals, plants or other organisms of an indigenous species cultivated, bred or kept in captivity or cultivated or altered in any way by means of biotechnology; any cultivar, variety, strain, derivative, hybrid or fertile version of any indigenous species; and any exotic animals, plants or other organisms whether gathered from the wild or accessed from any other source which, through the use of biotechnology, have been altered with any genetic material or chemical compound found in any indigenous species or any other animals, plants or other organisms. The indigenous biological resources covered by the International Treaty on Plant Genetic Resources for Food and Agriculture are, however, excluded.

Under the Act, permits are required for all bioprospecting projects and for the export of any indigenous biological resource to be used for bioprospecting or for any other kind of research. Stakeholders who provide access to resources or knowledge must be consulted, and their prior informed consent obtained before a permit will be issued.

The Act distinguishes between procedures to obtain indigenous biological resources, where an MTA is required between the applicant and stakeholder as well as a benefit-sharing agreement before a permit will be issued, and those to obtain knowledge, which require a benefit-sharing agreement. Ministerial approval is required for both MTAs and benefit-sharing agreements. Broad requirements are outlined in the Act for both benefit-sharing agreements and MTAs, and a Biodiversity Trust Fund is established into which all monies arising from bioprospecting projects must be paid. Benefit-sharing agreements must indicate the type and quantity of resources to be collected, the area of collection, the present and potential uses of the resources and the extent to which the stakeholders will share in the benefits. MTAs must set out the particulars of the provider and recipient, the type and quantity of resources to be provided, the area of collection, the purpose of export, potential use and conditions for transfer to a third party.

The procedures relating to the issuing of permits are not yet clear, with most of the detail left to be decided at a later stage in the implementation of the regulations. Other than for export purposes, research is excluded from the law. The requirement for the Minister to approve both benefit-sharing agreements and MTAs may prove to be a further impediment to facilitated exchange of material for research purposes, and there is some question as to whether this level of bureaucracy is necessary for the approval of MTAs, which simply cover the exchange of material between parties with a proviso not to commercialise them unless a separate benefit-sharing agreement is negotiated. The Government Gazette Vol. 472 published on 8 October 2004 indicates that the effective date for the implementation of Chapter 6 will be 1 January 2006, which will allow for the development of regulations and possibly the introduction of any changes that will facilitate the implementation of this Chapter.

11.4. The South African National Biodiversity Institute's mandate

In response to the International Agenda for Botanic Gardens in Conservation (www.bgci.org/policies/international_agenda.html) and the Global Strategy for Plant Conservation, the SANBI has developed a comprehensive Plant Conservation Strategy with measurable targets over a 2004–2010 time span. The ultimate long-term objective of these policy frameworks is to halt the continuing loss of plant diversity. As such the Plant Conservation Strategy forms a component of the Living Collections Policy of the SANBI. Within this policy, collection criteria are divided into three sections: horticultural, scientific and educational. The scientific component includes 'rare and endangered' plants as well as collections used for research purposes (horticultural, taxonomic and ecological).

Figure 20. Ismail Ebrahim and Rupert Coopman, from the Custodians of Rare and Endangered Wildflowers project based at SANBI, helping members of the Harmony Flats working group to identify and press plants occurring in their local municipal reserve. Less than 5% of this vegetation type (locally known as renosterveld) is left intact in South Africa, and thus local education and awareness programmes are vital for the future protection of these remnant patches of vegetation. About 180 plant species including eight Red Data List species are found on this site of only nine hectares, which is under severe threat from urbanization and invasive alien plant species (photo: SANBI).

11.5. The South African DNA bank in practice: implementation from the perspective of a developing country

For countries such as South Africa, which harbour extraordinary species diversity, the need for a national DNA banking facility is supported by the CBD, which recognizes national sovereignty over biological resources. As such our specific aim was to create a comprehensive centralized facility for plant DNA extraction and storage and to facilitate access to these DNA extracts for research on evolutionary biology, conservation and sustainable use, in line with the developing national legal framework. Through these efforts we have also aimed to alleviate the pressure placed on wild plant populations by making DNA extracts centrally available to researchers.

Using the funding made available by the Darwin Initiative, we were able to establish a plant collection programme specifically for the DNA bank, with the aim of archiving at least one representative species of the approximately 2,000 South African flowering plant genera by 2006. In addition to this, the Darwin Initiative funding covered purchase of an ultracentrifuge to enable purification of the DNA extracts for long-term storage and also recruitment of a full-time DNA bank manager to manage and curate the facility for a period of three years.

With the assistance and guidance of the Royal Botanic Gardens, Kew, we have been able to emulate the DNA banking facility housed in the Jodrell Laboratory at Kew and have benefited greatly from the training and technology transfer opportunities that have arisen during the course of this initiative. Perhaps most significantly, the South African DNA bank was jump-started by the duplication (and subsequent repatriation) of some 1,300 DNA extracts of South African plants held in the DNA bank at Kew. Under the terms of a Memorandum of Understanding (MoU) set out at the start of the Darwin Initiative funding cycle between SANBI and Kew (see Appendix 7), all South African DNA extracts will be reciprocally duplicated in the South African and Kew DNA banks. This will provide a 'back up' for each collection in the event that either suffers damage or destruction. The terms of the MoU also state that requests for DNA extracts of South African plants shall be directed to the South African DNA bank. Under these circumstances it is now possible to distribute DNA aliquots to interested research organisations under the terms of our MTA that is in accordance with South African legislation and access and benefit-sharing expectations.

An integral part of the initiative has also been to utilize the DNA bank to directly aid research in South Africa. We are capitalizing on our collection programme by endeavoring to produce a 'tree of life' of South African plant genera. It is anticipated that a phylogenetic classification of the flora will significantly aid bioprospecting and also allow measurements of biodiversity based upon DNA sequence data in addition to the identification of areas of endemicity for *in situ* conservation.

With respect to education and training, the DNA bank has provided an invaluable resource. The facility is being widely used by South African students enrolled in postgraduate studies, particularly those working in the fields of molecular systematics and conservation genetics. In addition, we have been able to establish short courses in molecular techniques, aimed at exposing young scientists in South Africa to DNA-based technologies and their application to biodiversity science. Perhaps most importantly, the South African DNA bank has provided a working example to young scientists of the important synergies required to carry out high quality and socially responsible biodiversity research. Firstly, it demonstrates the importance and requirements of well-curated collections (of both herbarium specimens and DNA extracts), both of which are needed to demonstrate reproducibility and authenticity in scientific endeavor; and secondly, the need to operate within the recommended framework of the CBD and national legislation – both within South Africa and elsewhere.

Figure 21. Ferozah Conrad, Angeline Khunou, Amelia Mabunda and Kholiwe Balele (DNA bank manager) from the Leslie Hill Molecular Systematics Laboratory at the Kirstenbosch Research Centre, SANBI (photo: SANBI).

In future we aim for the South African DNA bank to be extended to include rare and threatened species (particularly those threatened by illegal trade and where wildlife forensics may assist in law enforcement); more extensive population-level samples (again relevant to those species vulnerable to the wildlife trade); and in addition DNA and possibly tissue of other South African organisms (animals, fungi, microbes). In summary, our goal is to continue growing into a truly modern biodiversity institute – able to meet broad scientific and societal needs in line with new technologies and opportunities offered by the genetic era.

12. Tissue banking for DNA extraction at the Missouri Botanical Garden, USA
J. S. Miller

The last two decades have been a renaissance for phylogenetic studies, largely fuelled by the use of sequence data from a variety of genes to clarify patterns of relationship at various taxonomic levels. Researchers have studied an ever-increasing number of genes, and methods for amplification of DNA from multiple sources have improved. During this period, an increasing percentage of sequences have been obtained from herbarium specimens by removal of small pieces of tissue. Although DNA extraction methods continue to improve and the amount of tissue removed is small, herbarium curators are always justifiably concerned about removal and destruction of material from herbarium specimens that are intended to be permanent. Here, an outline is presented of the simple system used at the Missouri Botanical Garden (MBG) for the collection, storage and use of silica-dried leaf material as an alternative to removal of material from herbarium collections. MBG has not set up a proper bank of DNA extracts as described in this manual, but instead a collection of tissue samples for DNA extraction in other facilities outside MBG. This both protects herbarium specimens, satisfying the curatorial concerns of herbarium managers, and provides superior material to molecular researchers, ensuring a greater chance of successful amplification from the DNA samples. The system requires only minimal additional effort in the field and provides a collection or library of samples intended to be destructively sampled and used up to support molecular studies such as phylogenetics.

Most studies of relationships of higher taxa (including international programmes on assembling the Tree of Life and barcoding life) require access to plant material that represents multiple genera or species, usually from many different regions. Gathering material of large numbers of taxa from distant localities is difficult for a single researcher, so these studies are typically dependent on the supply of material from colleagues who are willing to collect, access to herbarium specimens or access to DNA collections. With a research programme active in many parts of the world, MBG was well positioned to collect and curate a collection of samples for DNA extraction that aims to be representative of higher-level taxonomic diversity, and garden botanists have gathered specimens that are representative of the diversity of families and genera throughout the world. As active systematists, they are aware of taxa of disputed taxonomic placement, poorly understood groups, the species that occur in

Figure 22. The Monsanto Center at the Missouri Botanical Garden houses part of the research programmes, including nearly six million herbarium specimens and the tissue banking facility for DNA extraction (photo: James Miller).

their areas that are likely to be of interest and those families and genera that are being actively studied, so the samples that they gather are often responsive to the needs of the molecular community.

The method for collection of samples is simple and involves collecting 10–20 cm^2 of leaf material in a ziplock plastic bag with about 30–40 g of silica gel. The size of sample bags (10 × 18 cm) was chosen specifically to fit into the storage cabinets that were selected to house the collection. An effort is made to collect young, actively growing leaves likely to be rich in nucleic acids (but be aware that unfortunately sometimes they also contain chemicals inhibiting PCR), to avoid any material that appears damaged or infected with disease pathogens and to maintain the samples out of direct sunlight during the period when leaves are being dried. If leaves are not dried in approximately 12–24 hours, the silica gel is replaced. Voucher specimens are collected with all tissue samples, and one of the duplicates is deposited at the MO herbarium so that it is available for consultation and confirmation of identity of the sample by anyone working with the DNA collection. Collectors record field notes as they would for any herbarium specimen, noting that there is an associated DNA collection, which triggers the production of an additional specimen label. Information on the collections is tracked in TROPICOS, MBG's fully web-available database (http://mobot.mobot.org/W3T/Search/vast.html).

a

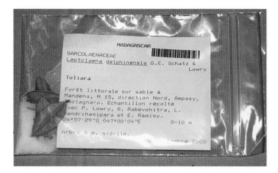

b

Figure 23. The Missouri Botanical Garden's tissue bank: (a) samples are maintained in a −20°C environment in file drawer cabinets that accommodate re-sealable plastic bags with desiccated leaf samples (b) (photo: James Miller).

Once samples have been shipped or otherwise transported and received in St. Louis, a portion of the silica gel is removed (to reduce necessary storage volume), but a small amount of silica gel is retained in each bag to ensure that leaves remain dry. A duplicate specimen label is inserted with the sample; it is then ready to be incorporated into the collection. The MBG collection is stored in a walk-in freezer maintained at −20°C and samples are stored in multi-drawer file cabinets.

All records for collections are maintained in MBG's TROPICOS database. An index to all DNA samples that have been accessioned is available at: www.mobot.org/MOBOT/research/diversity/dna_banking.htm. Each tissue sample has a duplicate specimen label prepared, and the sample label is simply barcoded (in the traditional sense of the word) and linked to the voucher record in TROPICOS. Barcode readers provide instant access to voucher specimen data and also the ability to print duplicate specimen labels that can be distributed with portions of any tissue sample. Thus each request for a sample can be accompanied by a duplicate specimen label produced by accessing TROPICOS through the sample barcode.

Most accessions in the collection are from collections made by Garden staff, associates or collaborators. However, MBG is willing to accept tissue collections made by other botanists who meet several criteria. Any material deposited in the tissue bank must be received with proper assurance that it was collected with all appropriate permits and approval of government agencies where collections were made. In addition, MBG is only willing to accept samples if a proper duplicate voucher specimen is included. Collectors who wish to deposit material in the Missouri collection must be willing to accept the restrictions outlined in our Material Transfer Agreement (see Appendix 5) and assign management responsibility for the collections to MBG.

The MBG collection was assembled specifically to support molecular systematics. A decision was actively made not to make the material available for any commercial research as most MBG botanists are collecting herbarium material with permits that only allow collection of material for academic research purposes. Securing permission from source countries to allow the samples to be used for commercial purposes would certainly be time consuming, difficult and probably expensive, so the decision was adopted that use of the collection would be solely for basic systematic research.

Each request to receive material from the tissue bank for subsequent DNA extraction is made via the MTA, which is available on the website (http://www.mobot.org/MOBOT/research/diversity/mta.pdf). The MTA provides an efficient means for informing users about restrictions on the use of the DNA samples. It specifically prohibits commercial use of the samples and requests written permission from MBG for the transfer of any material to third parties. It also requests that users appropriately acknowledge MBG, specific collectors and the countries that allowed export of the samples. In addition, the MTA allows MBG to efficiently track use of the collection by recording which samples were associated with each MTA in the files. MBG requests a fee of $25 to help defray the costs of the supplies for collection and curation of the tissue bank and the shipping costs to distribute the samples. These fees may be waived in cases where funds are not available; the aim of the fee is to help the collection operate and promote science, not to restrict access to the samples.

The collection currently contains more than 7,000 samples including representatives of 212 families of seed plants and an additional 26 families of mosses and ferns. The collection is particularly rich in Asteraceae (daisies) and includes significant collections from Chile, Ghana, Madagascar, New Caledonia, Panama, Paraguay, the United States of America and Zambia. It is hoped that this system will continue to take advantage of the active field programme at MBG. The regular fieldwork conducted provides an opportunity to continue to gather samples that could be an invaluable resource for molecular systematics.

13. Conservation genetics and the management of *ex situ* collections: examples from the Australian flora
S. L. Krauss

Biodiversity conservation is today concerned with much more than the management and protection of what remains through the establishment of protected areas. Increasingly, practical conservation efforts are focussed specifically on biological restoration or recovery. At the species level, the focus is on recovery of individual endangered species to reduce the threat of extinction and includes *ex situ* strategies such as DNA banking, seed storage, tissue culture storage and cryostorage. *Ex situ* conservation is a means to an end – not an end in itself – a tool for the reduction of immediate extinction risk and ultimately enhanced survival *in situ* through establishment of new populations (translocation or reintroduction into secure sites) or reinforcement of existing populations.

The global effort in *ex situ* conservation and *in situ* recovery through translocation will inevitably increase. For example, the 6th Conference of the Parties to the CBD adopted the Global Strategy for Plant Conservation (GSPC). This strategy has as a key objective to 'Improve long-term conservation, management and restoration of plant diversity, plant communities, and the associated habitats and ecosystems, *in situ* (both in more natural and in more managed environments), and, where necessary to complement *in situ* measures, *ex situ*, preferably in the country of origin'. In addressing this key objective, 16 outcome-oriented global targets were outlined for 2010, and target 8 of the GSPC specifically aims for '60 per cent of threatened plant species in accessible *ex situ* collections, preferably in the country of origin, and 10 per cent of them included in recovery and restoration programmes'.

Although *ex situ* and translocation conservation measures are the last line of defence and are often expensive, the need is clear. For example, in 1996, careful documentation by Coates and Atkins of the distribution of 1,386 populations of 'Declared Rare Flora' in Western Australia established that only 28% were on conservation reserves. The majority were on land not primarily set aside for conservation that included road-side reserves, private property and vacant Crown land with no designated use. Sixteen per cent of threatened species in Western Australia are known from only one population, 64% known from less than 1,000 individuals and 10% known from fewer than 50 individuals. Consequently, many populations are in highly vulnerable circumstances, for which *ex situ* conservation and translocation are critical for their preservation and recovery. In Western Australia, the Department of Conservation and the Botanic Gardens and Parks Authority (within which Kings Park and Botanic Gardens lies) have active programmes in *ex situ* conservation addressing this need.

The importance of *ex situ* conservation in protecting threatened species from extinction is demonstrated by the number of taxa that are maintained only in cultivation. In Hawaii, for example, at least six taxa are now thought to survive only in cultivation. Currently, over 10,000 threatened species are maintained in living collections throughout the world (botanic gardens, seed banks and tissue culture collections), representing some 30% of known threatened species. This could be increased to meet the proposed GSPC target by 2010, with additional resources and technology development and transfer, especially for species with recalcitrant seeds. Within this target it is suggested that priority be given to critically endangered species, for which a target of 90% should be attained. It is estimated that currently only 2% of threatened species are included in recovery and restoration programmes. Best practice in *ex situ* conservation includes careful genetic assessment, sampling and management.

13.1. Genetically representative sampling

An effective *ex situ* conservation project begins with the collection of a genetically appropriate and representative sample. Molecular markers are our best tools for the assessment of genetic variation within species. In the absence of genetic studies, guidelines exist for appropriate sampling. However, only in few cases do *ex situ* collections represent large and genetically diverse populations and they are typically descended from only a few founders. Poor sampling of remnant wild populations can have profound impacts on the outcome of re-introductions. For example, in a study from Rieseberg and Swensen in 1996, seed from only one or two maternal parents from the wild population of the Hawaiian endemic *Argyroxiphium sandwicense* subsp. *sandwicense* (Asteraceae) resulted in outplants with poor genetic representation, manifesting an atypical growth form and high levels of self-incompatibility. Genetic variation is thought to provide a buffer, enabling a species to survive environmental changes, which may be particularly important in *ex situ* conditions and for reintroduction into non-native sites. Genetic problems can arise in future generations from too limited genetic sampling, for example when: (i) inbreeding depression is associated with selfing; (ii) genetic drift leads to loss of potentially adaptive variation; or (iii) fixation of deleterious mutations occurs. In a worst-case scenario, low genetic variation can result in reproductive failure due to a lack of variation at self-incompatibility loci and populations may be sterile. At the other extreme, too broad a genetic sampling and the mixing of genetically distinct provenances can potentially lead to genetic problems, for example when outbreeding depression is associated with wide outcrossing, a reduction in the fitness of hybrids through the breakdown of coadapted gene complexes, genetic incompatibilities, or the disruption of meiosis through chromosomal differences.

The Corrigin grevillea, *Grevillea scapigera* (Proteaceae) is one of the world's rarest plant species. In 2004, less than five wild plants were known to exist in highly vulnerable native vegetation roadside remnants within the Western Australian wheatbelt – an area that has suffered extensive recent habitat loss through land clearing. Fortunately, *G. scapigera* has been the focus of intensive conservation efforts, predominantly by staff and students at Kings Park and Botanic Garden and volunteers from the wider community since the early 1990s. First collected in 1954, it was thought to be extinct in 1986 when the only known naturally occurring plant died. Since then, other plants have been discovered. In 1994, 27 plants were known to exist in eight highly vulnerable roadside verges. These and an additional 20 plants that were sampled for germplasm prior to 1986 were measured for genetic diversity by a DNA fingerprinting technique called random amplified polymorphic DNA (RAPD). Weak genetic differentiation among populations was found, suggesting that there was no provenance issue and populations could be mixed without negative genetic consequences. Total genetic variation was found to be high and comparable to more common congeneric species. Ten plants that encompassed 87% of all detectable genetic variation and displayed typical within-population levels of genetic variability were selected for micropropagation. Micropropagation was required because seed germination and propagation by cuttings were impossible at the time. However, micropropagation is costly and time-consuming, with success varying among genotypes. Therefore, careful selection was required to achieve maximum genetic variability efficiently. These ten genetically representative founders were used for translocation into two secure sites between 1996 and 1998, and in 1999, 266 plants of these ten clones were still surviving.

13.2. Assessment of genetic fidelity through recovery

Most genetic studies associated with *ex situ* conservation and/or translocation projects of rare plants have been limited to an assessment of genetic variation for founder population composition either before or after

a

b

Figure 24. (a) The Plant Conservation Centre, Kings Park and Botanic Garden, Perth, Western Australia. Opened in July 2005, this houses plant conservation research laboratories and off site conservation storage facilities; (b) *Grevillea scapigera* translocation site near Corrigin, Western Australia (photos: Siegfried Krauss).

translocation. However, maintenance of genetic fidelity, avoidance of genetic drift and/or prevention of genetic erosion through time in *ex situ* collections and/or translocated populations are important objectives that ideally require genetic monitoring.

In one of the few studies to assess genetic changes through a rare plant recovery programme, my colleagues and I used another DNA fingerprinting technique, amplified fragment length polymorphism (AFLP), to assess the genetic fidelity of the 266 plants of the Corrigin grevillea translocation mentioned above. Like most translocation projects, these plants were the product of many years of intensive propagation involving many researchers, community groups and volunteers. Each distinct genotype generated a unique DNA fingerprint. However, identical DNA fingerprints were generated for three differently labelled clones, indicating that these were mislabelled and are probably genetically identical (at least as far as the technique can measure). In addition, clone 27 planted in 1998 was genetically distinct from clone 27 planted in 1997, but was genetically identical to clone 33c. Propagation material of clone 33c was most likely incorrectly labelled clone 27 at some point between 1997 and 1998. All other clones were clearly genetically distinct, differing by between 24 and 52 band polymorphisms within their DNA fingerprints. Ultimately, eight clones, not ten, were present in *ex situ* collections at Kings Park and translocated populations. Perhaps more critically, because three differently labelled clones were in fact one, 54% (143 of 266) of all plants at the Corrigin translocation site were a single clone, and 90% of the population was made up of only four clones. Our result highlighted the difficulty of maintaining genetic fidelity through a large translocation programme.

13.3. Assessment of genetic erosion

A long-term objective for *ex situ* collections and self-sustaining translocated populations is the maintenance or increase of initial levels of genetic variation, which presumably reflects that found in wild populations. For outbreeders, large effective population sizes are typically required to avoid genetic decline caused by inbreeding and/or genetic drift and to maximise evolutionary flexibility to respond adaptively to environmental changes. Consequently, genetic monitoring of translocated populations is required to manage and prevent genetic erosion. Returning to the Corrigin grevillea, with the combination of new artificial techniques to germinate seed, the production of large numbers of naturally pollinated seed at the translocated population and an absence of natural seed germination *in situ*, seed was collected from these plants, germinated *ex situ*, and 161 seedlings were returned to the field site in winter 1999. We asked whether founding levels of genetic diversity were maintained in these offspring by using AFLP to (i) contrast genetic diversity in these offspring to their parents and (ii) assign paternity to offspring to assess the genetic contribution of parental clones.

We found significant genetic erosion in just one generation, such that the offspring were on average 22% more inbred than their parents and genetic variation declined by 20%. This rapid genetic erosion was largely a consequence of 85% of all seeds being the product of only four clones. Our result highlighted that rapid genetic decline may be a feature of many translocated populations when populations are small, which may ultimately threaten their long-term survival. We identified strategies to reverse this decline, which included equalizing founder numbers, adding new genotypes when discovered, optimising genetic structure and plant density to promote multiple siring and reduce kinship, promoting natural seed germination *in situ* rather than artificially germinating seeds *ex situ*, and creating a metapopulation of numerous translocated populations to restore historical distribution patterns and processes.

Genetic principles are being assessed in an experimental framework whilst achieving conservation outcomes in other translocation programmes at Kings Park and Botanic Garden. For example, the dioecious Western Australian endemic *Symonanthus bancroftii* (Solanaceae) was previously thought to be extinct. However, in 2004, a single male plant was re-discovered in the wild, and subsequently a female plant was also discovered. Material was collected and successfully micropropagated at Kings Park before the female plant in the wild died. These micropropagated plants were successfully hand-pollinated to produce a third genotype, a 'daughter'. These three genotypes are currently being used to establish translocated populations in secure sites within the species natural range. We are experimentally addressing the issue of whether it is better to establish translocated populations with the unrelated male and female genotypes rather than establishing populations with the father-daughter genotypes. If inbreeding depression is associated with father-daughter inbreeding, it may be better to focus initial efforts on the unrelated male and female individuals. Ultimately though, in the absence of the discovery of new wild plants, increasing effective population genetic size will require mating between parents and offspring or between full siblings. Genetic management of translocations of particularly rare species need to balance the genetic conflict between increasing population genetic sizes and minimising kinship with the more immediate concerns of maximising population size.

13.4. Maintenance of genetic fidelity under long-term storage

In modern plant conservation agencies, efficient *ex situ* preservation of large clonal collections for practical conservation outcomes is achieved through techniques such as tissue culture, slow growth (including cold storage) or cryostorage. These techniques are highly space efficient, minimise disease and pest problems and allow for manipulation and control of all external variables, which may cause irreplaceable loss of important mother plants when maintained outdoors. In particular, the technique of cryostorage optimises these advantages, has low maintenance costs and, theoretically, material can be maintained under these conditions indefinitely with virtually no maintenance.

However, for *in vitro* and cryogenically stored cultures, there may be risks to the maintenance of genetic fidelity associated with maintaining plant material under these conditions for extended periods of time. This may be due to the extended duration that these cultures have been maintained, as well as the use of various chemicals including plant growth regulators or cryoprotectants.

In one of the first studies to assess this risk in a conservation context, Turner and colleagues, in 2001, assessed genetic fidelity following tissue culture, cold storage and cryostorage for the threatened Western Australian endemic *Anigozanthos viridis* subsp. *terraspectans* (dwarf green kangaroo paw, Haemodoraceae). *Anigozanthos viridis* subsp. *terraspectans* is currently known from only four populations and is classified as rare and endangered due to extensive habitat clearing and limited distribution. As part of an integrated approach to its conservation, Kings Park and Botanic Garden has established a cryostored population of representative genotypes. Assessment of one of these genotypes following 12 months of storage revealed no loss of genetic fidelity as assessed by AFLP markers. The AFLP technique generated 95 markers, for which no qualitative differences could be detected among 45 samples from nine different cold-storage, cryostorage or tissue culture treatments.

13.5. Assessment of clonality

Assessment of extent of clonality within a threatened plant species is one of the most valuable practical conservation contributions from genetic analysis. An assessment of clonality identifies the number of genetically distinct individuals present, directly influencing sampling for *ex situ* conservation and strategies for *in situ* recovery. For example, identification of distinct genotypes would be critical for recovery of a self-incompatible rare species where artificial hand pollinations are proposed to restore reproductive success.

Grevillea pythara (Pythara grevillea) is a low (6–30 cm high) suckering, multi-stemmed shrub with characteristic black-tipped red flowers produced at the ends of branchlets. Discovered by a daughter of a local landowner on a degraded roadside remnant in the Western Australian wheat belt in 1990, *G. pythara* was declared Rare Flora in 1994 and ranked as Critically Endangered in 1995. The species is known from one location only, containing upwards of 100 'individuals' in three 'populations' along a roadside strip 900 m in length, and is thought to reproduce vegetatively from a single parent rootstock. The current distribution of *G. pythara* may have arisen from suckering of underground stems and subsequent disintegration of patches between current populations, or the plant may have been imported to the site in soil used in road construction works. *Grevillea pythara* appears to be sterile, as no fruits or swollen ovaries have been observed on flowers and pollen viability is less than 1%. Although hybrid origin for *G. pythara* has been suggested to account for its sterility, there appear to be no closely related taxa, and there is currently no firm suggestion as to possible parents for genetic testing. Genetic testing by AFLPs of 45 samples from all three 'populations' revealed that all samples were genetically identical.

The consequences of this genetic result are significant. For example, within the recovery plan, the criteria for success within the recovery criteria states that the number of individuals within populations and/or the number of populations have increased. Given the extent of clonality, unviable pollen and absence of fruit, it will be near impossible to increase the number of distinct genetic individuals above the current single individual. Clearly, the discovery of new populations is critical. Equally, these results indicate that the creation of new populations through translocation is critical for success under recovery criteria. Additionally, *ex situ* conservation of this genotype is critical, as the wild plant exists in a highly vulnerable situation in the wild. Fortunately, an *ex situ* collection has been successfully established by cuttings at Kings Park. Further genetic research is required to determine whether *G. pythara* is a sterile hybrid, which may influence the conservation effort invested.

Genetic assessment, sampling and management is a vital contribution to an integrated approach achieving practical outcomes in *ex situ* conservation and *in situ* recovery of rare and threatened flora. Although considerable attention has been devoted to genetic sampling issues for *ex situ* conservation, Husband and Campbell's 2004 survey of *ex situ* plantings and *in situ* reintroductions in the USA revealed that 85% of *ex situ* conservation programmes do not have explicit genetic criteria to guide them. Even more rarely have genetic criteria been assessed during storage or establishment of plantings *ex situ* and subsequent reintroduction efforts. The few examples discussed here have highlighted the important practical contribution that genetic analysis can play, not only in genetic sampling issues, but critically in management of *ex situ* collections and *in situ* reintroductions. When focussed on practical outcomes for conservation managers, genetic principles and management should be routinely applied as part of an integrated *ex situ* conservation programme.

14. The DNA bank of Brazilian flora in Rio de Janeiro Botanical Garden

L. O. Franco, M. A. Cardoso, S. R. S. Cardoso, A. S. Hemerly and P. C. G. Ferreira

Tropical ecosystems have suffered greatly in the last few hundred years as a result of human-mediated overexploitation. Extensive deforestation in particular has led to destruction of vast areas of tropical forest that form the natural habitat of many species, including plants. Among about 18 countries that account for about 70% of the plant and animal biodiversity known so far, Brazil is, most likely, the country with the largest biological diversity. The Atlantic rainforest is a hot spot of plant biodiversity, with a large number of endemic species. Most natural populations of these species comprise a small number of individuals and it is believed that for every two endangered Brazilian trees, one is found exclusively in the Atlantic rainforest ecosystem. As a result of proximity to urban centres, this complex ecosystem, one of the most threatened in the world, has been the subject of environmental degradation by social and economic pressures. During the last 500 years, the geographical range of the Atlantic rainforest has been reduced from 12% to 1% of the Brazilian territory.

Since 1994, a group of scientists comprising researchers from Rio de Janeiro Botanical Garden and the Medical Biochemistry Department of Rio de Janeiro's Federal University have been working in a multidisciplinary effort, with the goal of achieving a comprehensive view of the genetic structure, variability, phylogeny and systematics of selected endangered native species of the Atlantic rainforest. Besides the scientific importance, it is clear that a deeper knowledge of where and how the genetic resources are located in the remnant fragments of the Atlantic rainforest is of critical importance in setting up directives and actions of conservational public policies. Two years ago we realized that, because devastation of this biome is happening rapidly in spite of determined conservation efforts, a concerted initiative to gather, extract and preserve high quality DNA from plants and store this material in a consolidated collection, such as a DNA bank, would be crucial if a precise knowledge of the genetic resources still available was to be saved for present and future generations.

The Rio de Janeiro Botanical Garden DNA Bank aims to maintain genetic information representing the high diversity of the Brazilian flora. Our

Figure 25. The Rio de Janeiro Botanic Garden (photo: Luciana Franco).

collection targets include (i) Atlantic rainforest taxa, (ii) endangered and flagship Brazilian species, and (iii) selected species from our Arboretum and Herbarium collections.

We have sampled several representative plant species of the diverse ecosystems that shape the Atlantic rainforest biome *sensu lato*, including mangroves, sandy coastal plains, semi-deciduous seasonal forests, rainforests and high-elevation forests. The bank already contains about 200 specimens of species such as *Avicennia schaweriana*, *Laguncularia racemosa*,

Rhizophora mangle, Sideroxylon obtusifolium, Tabebuia cassinoides, Psychotria nuda, Psychotria brasiliensis and *Abutilon bedfordianum*. It is important to mention that, even if not endemic to these ecosystems, species of the most important Brazilian flora families, such as Poaceae, Fabaceae, Arecaceae, Myrtaceae, Solananceae, Euphorbiaceae, Apocynaceae, Annonaceae, Lauraceae, Cactaceae, Lecythidaceae, Malvaceae, Passifloraceae, Sapotaceae, Melastomataceae, Malphigiaceae, Sapindaceae, Meliaceae, Rutaceae, Bignoniaceae, Orchidaceae, Bromeliaceae and others, are also being included in the bank.

We have also banked multiple accessions of endangered and flagship Brazilian species collected in diverse regions where remnant populations can still be found. This will facilitate creation of a broad picture of their genetic structure. Here we can highlight 400 accessions of *Caesalpinia echinata* and approximately the same number for *Euterpe edulis*. The samples collected for these above mentioned targets come from expeditions conducted by our institution's researchers (field material is dried in silica gel before DNA extraction). All DNA samples are stored at -80°C with the equivalent voucher stored in the herbarium (RB), with the exception of species used in population genetics studies; in these cases only a few individuals per population are vouchered.

The Arboretum is essentially an *ex situ* conservation unit comprising thousands of specimens distributed in more than 200 plant families; DNA from selected species is being extracted and banked. Our main interest lies in the bromeliad collection. The Dimitri Sucre greenhouse contains the Brazilian Atlantic rainforest Bromeliaceae scientific collection with about 3,000 plants, dedicated to scientific research and conservation. Also of great importance is the legume collection, which comprises 200 species that are yet to be included in the bank, and the palm collection, with ten species already stored. In addition, we intend to include a native species component of the useful plants collection, including edible, medicinal, sacred and other non-timber species. Besides having an economic value, these plants are also important to the historical culture of Brazil.

The RB Herbarium, one of the oldest and most important in the country, was founded in 1890 and contains more than 500,000 specimens, including historical collections acquired in the 19th century by Pedro II, Emperor of Brazil. We have been obtaining encouraging results with DNA extraction from herbarium specimens, with 30-year-old samples shown to yield DNA of sufficient quality to be amplified using standard PCR techniques. Our efforts have been concentrated in families that are being used for phylogenetic studies by our institution's researchers, such as Melastomataceae.

Our main problem in establishing the DNA bank was, and still is, a combination of reduced economic resources for expeditions, reduced IT support, difficulties in importing reagents and equipment and a small number of taxonomists when compared with the great plant diversity found in Brazil; of all active taxonomists in the world today, only 6% live and work in Latin America. Even though this number is increasing, it is still insufficient when compared to the richness of the Brazilian flora. Therefore we are facing the fascinating task of establishing a high profile DNA bank with our limited resources in a country recognized as having one of the highest plant diversities in the world.

In the short and long term, the DNA bank of Brazilian flora will be a source of genetic material for research, laying the essential groundwork for conservation and biotechnology. DNA samples stored in the bank are available to researchers all over the world, dependent upon their institution developing projects in collaboration with a Brazilian public institution. The Brazilian institution is responsible for requesting authorisation for access and shipment of the material in accordance with Brazilian laws formulated upon the basis of the Convention on Biological Diversity. The Rio de Janeiro Botanical Garden DNA Bank is sponsored by Aliança do Brasil – Companhia de Seguros.

Figure 26. DNA banking facility at the Rio de Janeiro Botanic Garden (photo: Luciana Franco).

15. Conservation genetics and population-level banking: the United Kingdom DNA bank as a case study

M. F. Fay, R. S. Cowan, I. Taylor and J. Sutcliffe

In parallel with DNA banking for phylogenetic purposes (whereby only one or a few samples per species are normally accessioned), many institutions now have collections of DNA sampled at the population level, that is, geographic sampling with multiple samples per population and multiple populations per species. This approach to sampling allows a different set of questions to be addressed. Examples of these questions are: (i) how many species are there? (ii) are populations distinct? (iii) which populations are the most variable? (iv) is there gene flow between populations? (v) which populations should be targeted as high priority for *in situ* conservation or *ex situ* seed banking in the case of plants? (vi) which taxa are native? (vii) which taxa should be used for re-introduction and/or restoration? and (viii) has the genetic structure of populations changed over time?

For the last ten years, the Royal Botanic Gardens, Kew, in addition to its global DNA bank, has been archiving DNAs of individual samples from plant populations, mostly for conservation purposes. Samples to be included in this population-level DNA bank are selected through discussion with representatives of the conservation bodies commissioning the study, for instance statutory nature conservation agencies, including English Nature and the Countryside Council for Wales. As of 2005, most of the species included in this DNA bank are rare (many are endangered or critically endangered), and thus these are regarded as high priorities for genetic studies to enable better conservation management. Some common species have also been extensively sampled to allow comparison between common and rare species. An example of the former is *Orchis mascula*, one of the most common orchids in Britain. Genetic studies have been conducted on this species to allow comparisons with its close relatives, *O. purpurea*, *O. militaris* and *O. simia*, all of which are much rarer (the first is locally common and the last two are extremely rare in south-east England). Population samples can also be used to study species delimitation in complexes of closely related species: examples of studies carried out at Kew have included *Dactylorhiza*, *Limonium* and *Sorbus*. For these and other genera with difficult species complexes, one of the main priorities has been to better understand species delimitation so that appropriate conservation goals can be incorporated into subsequent action plans.

15.1. Sampling

Issues relating to the level of sampling required for genetic studies have been long debated, and we advocate a pragmatic approach. Where populations are small (<20 plants) we normally collect material from as many plants (ideally all plants) as we are permitted to, but for larger populations, we normally collect about 20 samples per population. For rare species, however, a degree of sensitivity is required, and this often involves collection of fewer samples than would normally be called for in population genetics theory. There is no benefit in conducting the study if it results in damage to the population(s) or a breakdown in relations between geneticists and conservation managers.

Most samples are collected into silica gel using standard methods, but in some cases DNA is extracted from freshly collected material. Both approaches yield high quality DNA if appropriate material (such as non-senescent leaves or flowers) is collected. Surprisingly small samples are necessary in many cases. For example, with many orchid species (*Orchis* or *Dactylorhiza*), one or two flowers yield sufficient DNA for our purposes. For small plants, such as terrestrial orchids, collecting a small number of flowers from a plant with many flowers is likely to be less damaging than collection of leaf material.

15.2. Vouchers

Taking material for vouchers requires pragmatism. We aim to collect one voucher per population (not one per individual), but it is not possible in all cases to collect either a full voucher or even a small amount of extra material. For example, permission to collect material for DNA extraction from the lady's slipper orchid (*Cypripedium calceolus*) in England was granted, but collection of a herbarium voucher to accompany the DNA samples was out of the question. Photographic e-vouchers provide a good alternative. Single flowers of many orchids have been preserved in standard fixatives to act as vouchers. Normally, the absence of a voucher for each plant in a population is not problematic. If genetic data subsequently raise questions about identification or introgression, sites can be revisited and the situation reassessed as necessary.

15.3. DNA extraction

Only small quantities of DNA are required for most modern techniques, and this means that these techniques can be applied to rare species for which only small amounts of plant material can be collected. Storage (banking) of the resulting DNA also means that re-sampling is not necessary because sufficient DNA can be extracted for multiple studies from a single sample. There are many methods of DNA extraction available and choices may have to be made, balancing expense against quality of the end product. At Kew, our 'gold standard' is the modified 2× CTAB method followed by purification on caesium chloride gradients. This method is used for the most important samples, but limitations in terms of cost and time mean that this cannot be used for all the population samples maintained at Kew. Our two primary alternatives are to store the DNA in Tris EDTA buffer as per the 2× CTAB method without continuing through the caesium chloride gradient, and/or to clean an aliquot of the DNA with a proprietary column system (e.g. Qiagen) and store the cleaned product.

It is likely that DNA cleaned using these quicker and cheaper techniques will be more subject to degradation over time than that from caesium chloride gradients. The choice of method needs to be taken on a case-by-case basis: if plant material is unlikely to be available again, we normally opt for gradient purification, but if material is not likely to be a limiting factor, one of the other techniques is used.

15.4. Genetic studies

A wide range of techniques is now available, and decisions relating to which technique to use are based on a number of considerations, including: (i) question or questions being asked; (ii) time available; (iii) cost; and (iv) biology of the species.

For many questions, including assessment of species delimitation, the degree of genetic variation within and between populations and degrees of hybridisation/introgression, the amplified fragment length polymorphism (AFLP) technique has proved to be a good tool in many cases. British examples include hybridisation among *Schoenoplectus* (Cyperaceae) and *Sorbus* (Rosaceae), origin of allopolyploids in the orchid *Dactylorhiza* and species delimitation in *Sagina* (Caryophyllaceae).

Advantages to the AFLP technique include its ready applicability to new species (it does not require much development for each new case) and that it is relatively robust in terms of reproducibility. Disadvantages include an inability to distinguish between homozygotes at a particular locus in most cases (and therefore limited application of standard population genetics statistics) and that it is not readily applicable to species with large genomes. Some of the terrestrial orchids that are rare in Britain have large genomes (e.g. *Cephalanthera* spp., *Cypripedium calceolus* and *Orchis* spp.), and this has hampered our use of AFLP markers in these cases. Large genomes are relatively widespread in monocots, but are rare in eudicots

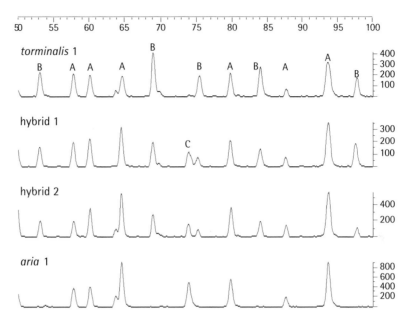

Figure 27. Specimen traces for samples of *Sorbus aria* (whitebeam) and *S. torminalis* (wild service tree) and hybrids between them, showing a generally high degree of similarity and also additivity in the hybrids. Examples of different types of band are indicated by different letters: A = universal bands, B = bands present in *S. torminalis* + hybrids, C = bands present in *S. aria* and hybrids. Numbers at the top of the figure indicate the sizes of the fragments in base pairs and numbers at the side indicate the strength of the peaks in arbitrary units of fluorescence (photo: Mike Fay).

(the only known cases being *Viscum* spp.). In the flora of the British Isles, the main groups in which this is likely to be a problem are terrestrial orchids and members of Alismataceae and Liliaceae.

Nuclear microsatellites, another DNA fingerprinting technique, are increasingly used because they provide information on levels of heterozygosity and the resulting data can therefore be analysed in similar ways to enzyme data. They do, however, have to be developed for each new species or group of species studied, and this is relatively expensive and time consuming. New techniques for their development are becoming available and some of these streamline the process. As more sets of nuclear microsatellites are developed for different groups, the problems of development time and cost will diminish because, in many cases, loci will have been developed for related taxa. For example, it has already been shown that some microsatellite loci developed for *Malus* (apples) can be used in *Sorbus*. Thus, the expense of development was borne by the agricultural industry, and these markers can now be applied to conservation genetic questions.

Box 11 Kew's DNA bank V. Savolainen

Kew's DNA bank was established in 1993 when Professor Mark Chase took up the position of Head of Molecular Systematics within the Jodrell Laboratory. Just over a decade later, the DNA bank contains over 25,000 samples of plant genomic DNA, representing all major plant groups (about 5,000 genera from nearly all angiosperm families), all stored at -80°C. The DNA bank database is available online (www.rbgkew.org.uk/data/dnabank) and samples can be ordered directly from the web. The fees charged do not imply that DNA samples are being purchased; rather, this fee is meant to offset the cost of producing the sample and maintaining and shipping it. Kew's standard terms of supply (see MTA in Appendix 2) prohibit commercial application of the DNA supplied. The great majority of these DNA samples are of high molecular weight and reasonable concentration. Some, as noted in the comments column on the online database, are somewhat degraded because the tissue from which they were extracted was not dried properly, whereas others are more dilute than desirable, but some taxa never produce much DNA. Several thousand of the DNA samples have been extracted from fresh material growing in the Gardens (Kew's living collection has about 30,000 species – around 10% of all known angiosperms), but the rest have come from research projects hosted in the Jodrell Laboratory, sometimes involving DNA extractions from old herbarium sheets. DNA samples at Kew have been cleaned on EtBr/CsCl gradients (see Appendix 1), so they are reasonably free of RNA and *Taq* enzyme-inhibitory chemistry, but in some cases further cleaning and concentration by means of a silica-based column may be necessary before PCR amplification can be achieved. The amount of DNA sent externally by Kew (typically about 25 µl) is suitable for PCR-based techniques; if researchers require a greater amount of DNA, this may be possible, but they then should contact one of the two full-time DNA bank managers.

Figure 28. The Palm House at the Royal Botanic Gardens, Kew, UK. As of 2005, Kew hosts the largest DNA bank for plants in the world (photo: RBG Kew).

Plastid microsatellites only provide information about one parent (normally maternal in angiosperms), but the lack of recombination of these markers makes them useful in the study of biogeographical patterns. They also give a measure of genetic diversity and have been used in *Cephalanthera*, *Cypripedium*, *Orchis* and *Sorbus* at Kew, either to avoid the problems with AFLP markers due to genome size (the orchids) or to provide an additional data set (*Sorbus*), thus providing answers to different questions and corroboration of general patterns of variation.

DNA sequencing also has some application in genetic studies for conservation. We are, for example, currently using sequences of the internal transcribed spacer of nuclear ribosomal DNA (ITS) to investigate levels of hybridisation/introgression in *Orchis* spp.

New techniques are still being developed, and different DNA-based markers may become available in the future (e.g. DNA barcodes, Chapter 3). Some of these are likely to be applicable to conservation studies. In the early 1990s, random amplified polymorphic DNA (RAPD), for example, was widely used, but this technique has now largely been replaced by more sensitive and reliable techniques including AFLP. The literature on techniques developed for crop plants should be watched closely for advances that may potentially be applied in wild plants.

15.5. Outcomes for conservation

Genetic studies are now being used in many cases to inform conservation management decisions. In the past, conservation decisions were normally taken without genetic data to support them and were based on criteria such as size and geographical distribution of populations. Genetic data do not replace these sources, but they do add considerably to our understanding. To illustrate this, we will use the example of *Orchis militaris* in England. Previously thought extinct, three populations were discovered in the mid-20th century. Two of these (A, B) are close to each other (9 km) and the third (C) is approximately 200 km away. Two (A, C) are relatively large (>200 individuals), whereas B consisted of only six individuals when it was discovered. Its small size and proximity to A led people to believe that B was a recent expansion from A, and although this was clearly good news, it was considered to be genetically part of A. Thus, the two important populations were deemed to be A and C. However, genetic studies using AFLP and plastid microsatellites have demonstrated that all three populations are distinct from each other, and that B is not derived from A. In addition, all six plants in B are genetically distinct from each other and the level of variation is similar to that found in A. In stark contrast, C was shown to possess little genetic diversity, and this population is likely to be the result of a relatively recent migration from the continent of a small number of seeds, which have given rise over a number of generations to the current population. Loss of a single plant at B would represent a significant loss of genetic diversity, not only at that site but also for the species in England, whereas loss of a larger number of individuals at C would probably not result in loss of a significant amount of genetic diversity. In this case, population size and proximity were misleading about the expected levels of variation in each population, and genetic studies completely revised conservation priorities for this species.

Increasingly, DNA banking and associated genetic studies are providing useful tools for conservation projects, and data from such studies are being incorporated into management plans for rare species. In addition to population-level data, information relating to species delimitation and levels of hybridisation/introgression is also a useful outcome in many cases. In our experience, DNA banks for conservation purposes can be run in parallel to those for phylogenetic purposes, but different considerations need to be taken into account regarding sampling, vouchering and DNA purification methods. Appropriate decisions will make these DNA banks valuable resources for many years to come.

16. The Ambrose Monell Cryo-Collection: a museum-based molecular collection
A. Corthals

At the beginning of the 21st century, natural history collections are facing the challenge of keeping up with scientific investigation into the nature of life and biodiversity at the molecular level. Historically, most biological materials in natural history collections have been dried or formalin-fixed and have not often preserved nucleic acids in workable quantities or of sufficient quality to perform genetic research. The preservation and long-term storage of biological specimens, derived materials (including DNA extracts) and associated information is essential to ensure comparability and reproducibility in all areas of biological research. At the American Museum of Natural History, as in most academic institutions around the world, researchers working prior to 2001 typically collected frozen tissues as part of individual research projects, with resulting biomaterials stored within individual departmental laboratories. The need for a centralized repository stems from the eclectic history of acquisition, dispersed storage and inconsistent curatorial systems. Storage problems reflect lack of appropriate knowledge and resources for handling and maintaining tissue samples in departments focused on the care and maintenance of large collections of traditional specimens (e.g. whole organisms, skulls, skins and skeletons). With time, equipment and protocols change for the better, but materials used and specimens collected in the past are in serious jeopardy of deterioration if they are not included in projects to retroactively upgrade their curation.

To address these issues, the American Museum of Natural History, in partnership with the US National Aeronautics and Space Administration (NASA) and major private support, established in 2001 a new collection unit dedicated to the cryopreservation of tissue specimens, the Ambrose Monell Cryo-Collection (AM-CC), for genomics research on the Earth's biodiversity. Since its inception, the AM-CC has played a key role in helping to establish national standards for such super-cold tissue collections.

16.1. AM-CC mission

Our mission is to provide an accessible repository of frozen tissue specimens, collected and maintained under rigorously controlled conditions. In a time of massive species loss, such efforts are essential to preserve a record as comprehensive as possible of the Earth's biodiversity.

The broad scope of the AM-CC addresses an under-served niche within the cryogenic biorepository community by attempting to catalogue all biodiversity at the molecular genetic level. The AM-CC is further distinguished from other repositories because it exists within the framework of the American Museum of Natural History, where tissue samples can be referenced with documented collecting events involving traditional voucher specimens and associated data. Here, modern bioinformatics initiatives will ultimately link collections with taxonomic determinations, bibliographic citations, geospatial referencing information, genetic data, digital images and photographs.

The AM-CC supports a broad range of comparative genetic and genomic research initiatives. We provide our researchers with collecting kits to readily sample and ship genetic material of high quality, enhancing the genetic information content of each specimen. The AM-CC supports ongoing genetic research by ensuring that all research materials are vouchered. We extend this service to the entire scientific community, under the guidance of our institutional policy (http://research.amnh.org/amcc/coll_pol.html).

Scientists using the AM-CC have access to legally collected, authoritatively identified and properly documented specimens, complete with museum catalogue numbers to reference in their scholarly publications. In many cases, each tissue specimen is linked through its catalogue number to a morphological voucher maintained in one of the traditional departmental collections. However, many specimens are harvested from living animals (both captive-born and wild) and have only a tissue voucher, whereas still other accessions simply consist of nucleic acid extracts. In each case, we record as much information as possible to document the sample: where and how it was collected and by whom, as well as how taxonomic identification of the specimen was determined and by whom, and what research has previously been conducted on the specimen.

16.2. System and equipment

The AM-CC maintains specimens in an array of liquid nitrogen cooled vats at temperatures below -150°C. Frozen specimens held at -20°C can be subject to significant protein and lipid changes and damage from the growth of micro-organisms; specimens held at -80°C can also be subject to protein and lipid changes, with extensive desiccation of specimens and some molecular damage. Thus, we advocate a 'colder is better' position on archiving tissues, especially when the long-term use of the resource is undefined.

a

Figure 29. (a) Voucher drawer at the Ambrose Monell Cryo-Collection, showing the specimens with a barcoded ID label (b). The barcode used is the same as for the tissue extracted from the voucher and links both tissue and voucher records, while giving them different positions within the collection (photos: Julie Feinstein).

b

Figure 30. The AM-CC cryostorage room at the American Museum of Natural History (photo: Angélique Corthals).

The AM-CC lab facilities include a dry lab, complete with two biosafety cabinets (where the accessioning of the samples is mostly carried out), a fully equipped wet lab for all kinds of genetic assays and a cryostorage room, in which the physical collection resides. Each sample is archived in a vial to which a cryo-resistant label is applied containing the AM-CC unique barcode and human-readable number.

Additionally, the AM-CC has a wide array of equipment for field collection trips, such as our cryogenic dry shippers, which facilitate freezing samples collected in the field and transporting them back to the museum without thawing the specimens. We also put digital cameras at the disposal of AMNH researchers to encourage the capture of e-vouchers.

16.3. The problem of non-centralized repositories

We strongly advocate the use of a centralized repository as a way to solve problems of security, safety and quality control. Indeed, un-gated and multiple tissue collections within the same institution, such as is the case in many museums and universities around the world, face three main problems jeopardizing their collections.

The first problem involves inadequate or non-standardized containers and labelling; a variety of containers and labelling techniques are used for making tissue collections. Many of the containers prove to be inadequate for long-term archival use. In particular, vials with external threading are susceptible to cracks and loss of air-tightness, eventually leading to desiccation and oxidation of specimens and/or leakage. Moreover, many of the techniques for labelling are proving to be inadequate: pieces of tape that were applied to collection vessels do not adhere properly under long-term storage conditions at low temperatures. In other cases, the writing itself is disintegrating with changes in temperature associated with specimen sorting and retrieval and/or buffer leakage.

Secondly, many tissue collections are located in mechanical freezers without sufficient back-up freezer space to handle meltdown of a malfunctioning freezer or back-up power in the case of extended service interruptions. Past power failures, such as the blackout of August 2003 in the USA, have revealed the urgency of accessioning samples in reliable, non-mechanical, non-power based freezers, such as the cryo-vats of the AM-CC. Furthermore, tissue collections with unrestricted and unguarded access to the freezers run a high risk of disappearance, accidental thawing (if samples are left out by mistake), contamination and misplacement of specimens. In most cases, specimens to be used for research or sent out on loans are sub-sampled within laboratories that house numerous PCR machines and no biosafety cabinets. This severely increases the risk of contamination of the samples through aerosol DNA, as shown by Scherczinger et al. (1999).

Thirdly, electronic databases for many tissue collections lack location references for the freezers. For most collections, the retrieval of samples relies critically on the memory of the collection manager and a single tabulator spreadsheet to locate specimens.

As a 'biosafety II'-restricted centralized repository, the AM-CC has answered these numerous problems both through the design of facilities and through bio-informatics solutions. We maintain samples in stable, liquid nitrogen charged cryogenic freezers. The AM-CC also has ample back-up freezer space in case of an emergency. Furthermore, not only does the AM-CC have a restricted access policy and gate-keepers for the collection, but contamination is also avoided by the use of two biosafety cabinets in which all transfers are performed and absence of PCR machines within the facility. All instruments are also rigorously disinfected between the transfer of each specimen and the biosafety cabinets are sterilized at the end of each day.

16.4. AM-CC databasing and inventory

As part of daily operations, the AM-CC tissue samples are indexed using Freezerworks, a relational database application program well-suited to the task of freezer inventory management. The program creates a record for each specimen, giving it a unique barcode identification. Data entry in the database is made easy and reliable with an 'import' feature, allowing donors to generate their own data spreadsheet, which is then imported into the AM-CC database without a third party (such as a lab technician) modifying anything manually. Each record contains data ranging from the collection event (who collected it, when, where, how and even why) to physical characteristics of the donor organism and position of each vial in the collection's many freezers. Following data entry, the program generates a printed cryo-resistant label, which includes the unique identification number both as a 'traditional' barcode as well as a human-readable numeric string. This feature allows lab technicians to retrieve any vial from the freezers quickly and reliably by scanning the label for its associated data and thus confirm the identity of the specimen they are attempting to retrieve. The computer database tracks each barcoded vial, noting the specimen's taxonomic identity, where the specimen was collected and by whom and how many times the sample has been thawed and refrozen.

Because Freezerworks has not yet been made compatible with the Internet, a 'shadow' database has been built to render the holdings of our collection accessible to the scientific community worldwide. To host the fully searchable database, the AM-CC launched its own website in 2002: http://research.amnh.org/amcc.

The data available on the AM-CC website constitute a subset of data from the Freezerworks database. This allows the AM-CC to have better control of the amount of data published on the web. The website database is updated every three months during which time the entire data set is exported from Freezerworks to a text file. Once checked using DataCheck software and quality-controlled, the data set is ready for the last steps of web interface. The on-line database of the AM-CC runs on the MySQL relational database management system (see Figure 15). This database allows specimen records in the collection to be located by taxon name or browsed by taxonomic hierarchy. The web database front-end is written in PHP and is served on the back-end by an Apache web server running on Solaris. The PHP code is in a development stage but is being steered towards an object-oriented design that should make it fairly portable to other collections.

The database is designed to integrate with the National Center for Biotechnology Information's (NCBI) Entrez indexing and retrieval engine. This allows for AM-CC records with nucleotide sequence accession numbers to link to corresponding pages on the NCBI Genbank and Taxonomy databases and inversely for GenBank sequences to link back to the AM-CC.

The core tables of the database are the specimen and taxonomy tables, which are joined by an associative table that allows a many-to-many relationship between specimens and taxonomic names. In addition, both the specimen and taxonomy tables form a many-to-one relationship with a table for 'foreign resources', which flexibly stores information about external URLs, files or database resources. As of 2004, only website URLs are stored in these tables, but the design, in theory, allows the storage of any type of resource-location information pertaining to an address that can be associated with an institution. In the future, the taxonomic names will be joined many-to-many with a common names table.

16.5. E-Vouchers in the AM-CC

An e-voucher is a digital representation of a specimen and has special utility in the case of frozen repositories, where it is often impractical or even impossible to collect and store traditional vouchers (usually the cadaver of a whole animal). To that effect, the AM-CC has linked, whenever possible, the specimen records to digital images (hosted by the museum's digital library server), making a complete connection between sequence data and the visual identity of the specimen examined.

Appendix 1: Isolation of total plant cellular DNA for long-term storage: CTAB procedure

L. Csiba and M. P. Powell

The DNA extraction protocol outlined below is a modified version of Doyle and Doyle's method (1987).

DNA Extraction

1. Preheat 10 ml 2× CTAB buffer containing 40 µl beta-mercaptoethanol in a 50 ml centrifuge tube in a 65°C water bath. If using fresh, silica-dried or herbarium material, preheat a mortar and pestle to 65°C. If using frozen tissue, cool a mortar and pestle to 4°C.

2a. Grind 0.5–1.5 g fresh leaf tissue in a preheated mortar and pestle with a small portion of the buffer. For tough tissues a pinch of sterile sand or silica gel may be added. Add the remainder of the buffer to the mortar and swirl to suspend the slurry. Pour slurry into the 50 ml tube and return the tube to the 65°C water bath and incubate for 15–20 minutes with optional occasional gentle swirling and mixing.

or

2b. Grind 0.3–1.5 g frozen leaf tissue with liquid nitrogen in a pre-cooled mortar and pestle. Transfer the powdered sample to the 50 ml tube containing the buffer. Swirl gently to break up the frozen lump and return the tube immediately to the 65°C water bath. Incubate for 15–20 minutes as above.

or

2c. Grind 0.3 g silica-dried leaf material or 0.2 g herbarium material in a preheated mortar and pestle. Add a small portion of the buffer and grind until a uniform slurry is obtained. Add the remainder of the buffer to the mortar, swirl to suspend the slurry and pour it into the 50 ml tube. Return the tube to the 65°C water bath and incubate for 15–20 minutes as above.

3. Add 10 ml SEVAG, mix gently but thoroughly. Open the cap of the tube to release gas, then re-tighten and rock the tube gently using an orbital shaker (100–150 rpm) for up to one hour. This may take up to one and a half hours for slimy or mucilaginous samples.

4. Centrifuge at 8,000 rpm for 10 minutes at 25°C. Ideally, the aqueous (top) phase, which contains the DNA, will be clear and colourless.

5. Remove the aqueous phase with a plastic transfer pipette and transfer to a 50 ml falcon tube. Dispose of SEVAG and plant debris in a designated SEVAG waste container. Do not over-fill the waste container above its shoulder.

6. Estimate the volume of the aqueous phase (typically c. 20 ml) and then add 2 × this volume -20°C ethanol (for fresh, frozen and silica-dried specimens) or $^2/_3$ of the volume -20°C isopropanol (for herbarium specimens) and mix gently to aid precipitation of DNA. Put in -20°C freezer to precipitate DNA; as a guideline, leave silica-dried and fresh material for 1-2 days and herbarium material for about two weeks (see note 3).

Density gradient

7. Centrifuge at 3,200 rpm for 5 minutes to collect precipitate. Pour off liquid and add 3 ml 70% ethanol to wash the pellet. Dislodge the pellet to facilitate 'washing' and then leave standing for 15 minutes.

8. Centrifuge at 3,200 rpm for 3 minutes. Pour off liquid and drain upside down for 5–10 minutes to allow alcohol to evaporate. If the pellet appears loose it may be better to lay the tube on its side. Leave the tubes in a fume cupboard overnight to allow the alcohol to evaporate completely.

9. Resuspend DNA in 3 ml ethidium bromide/caesium chloride (EtBr/CsCl) solution. Cover samples in foil to keep dark (EtBr degrades in sunlight). Leave in a shaker overnight or for a few days, until the pellet dissolves.

10. Pour re-suspended DNA into ultracentrifuge rotor tubes. Add EtBr/CsCl until the total weight of the tube is 8.04–8.06 g.

11. Spin at 45,000 rpm for at least 12 hours or 58,000 rpm for 5 hours in an ultracentrifuge. The first option is preferable as it places less pressure on the tubes within the ultracentrifuge, but the second option can be used if time is a limiting factor.

12. Remove tubes from the ultracentrifuge and place them on an ultraviolet trans-illuminator. The DNA should be visible as a distinct band under UV light. Remove 1,200 ml of the band and transfer to 5 ml tubes (sometimes it may be necessary to remove some of the solution above the band to prevent spillage). If it is not possible to see the band, 1,200 ml should be removed at the same level as the band in the other samples. In this instance, the rest of the sample should be saved until the final gel check of DNA has shown that the portion removed did contain the DNA.

Dialysis

13. Add 1,200 ml of butanol saturated with sodium citrate, shake the tubes and leave them on their side for 15 minutes to remove EtBr, which separates into a distinct layer. Shake tubes gently occasionally.

14. Rinse a 5 L beaker and fill to 4 L with de-ionised water (2 L if fewer than 12 samples).

15. Cut dialysis tubing into 10 cm strips and rinse in de-ionised water.

16. Clamp the lower end of each piece of tubing with a dialysis clip and pipette samples (lower layer) into tubing. Clamp the top end of tubing, taking care to exclude air bubbles from the sample.

17. Place samples in the 5 L beaker, making sure the samples are kept under water. Leave for 4 hours on a magnetic stirrer.

18. Transfer the samples to a tray and cover with sugar. This will concentrate the DNA. Leave samples for 20–40 min depending on the degree of concentration required to obtain about 1 ml DNA.

19. Rinse out the 5 L beaker and add 50 ml dialysis buffer (80× TE buffer) and 4 L de-ionised water (or 25 ml buffer and 2 L water for 12 samples or fewer). Put samples into the beaker and leave for a minimum of 4 hours (overnight if necessary), again ensuring that they are kept under water.

20. Remove the samples from the beaker and repeat step 19.

21. Transfer samples into 1.5 ml microcentrifuge tubes. Check DNA levels on an agarose gel and store at -20°C or -80°C for long-term storage.

This method typically yields 20–100 mg high molecular weight DNA from 0.5-2 g fresh (or fresh frozen) leaves.

Notes

1. We find that higher yields of total cellular DNA are often obtained from species with succulent tissues, particularly orchids, by increasing the CTAB buffer from 2 × to 3 × concentration. Many succulents also require the addition of 1–2% PVP (poly vinyl pyrrolidone) to the CTAB buffer.

2. Fibrous leaves are often homogenized more efficiently by grinding in liquid nitrogen (either with mortar and pestle or in a coffee mill), followed by addition of CTAB buffer, rather than directly in the buffer.

3. Care should be taken to ensure that DNA samples are not left precipitating in alcohol for longer than 2-3 months. If this does occur, then products which inhibit PCR reactions are likely to be co-precipitated and render the DNA sample unusable for future experiments.

Buffers

2× CTAB buffer:

100mM Tris HCl pH 8.0
1.4M NaCl
20mM EDTA
2% CTAB

SEVAG:

24:1 chloroform:isoamyl alcohol

TE buffer:

10mM Tris HCl pH8
0.25mM EDTA

EtBr/CsCl:

1.55 g/ml

Appendix 2: Royal Botanic Gardens, Kew Material Supply Agreement for DNA

 NON COMMERCIAL
MATERIAL SUPPLY AGREEMENT FOR DNA
(with effect from 1 November 2004)

The Royal Botanic Gardens, Kew ("Kew") is committed to the letter and spirit of the Convention on Biological Diversity ("CBD") and expects its partners to act in a manner consistent with the CBD. This agreement is designed to promote scientific research and exchange, whilst recognising the terms on which Kew acquired the plant or fungal material and the important role played by *ex situ* collections in the implementation of the CBD. Kew reserves the right not to supply any plant or fungal material if such supply would be contrary to any terms attached to the material and/or to the CBD.

Kew will supply the material listed on the reverse of this agreement ("Material") subject to the following terms and conditions:

1. The recipient may only use the Material, its progeny or derivatives for the common good in **scientific research, education, conservation and the development of botanic gardens**;

2. The recipient shall **not sell, distribute or use for profit or any other commercial application**[1] the Material, its progeny or derivatives;

3. The recipient shall **share fairly and equitably** the benefits arising from their use of the Material, its progeny or derivatives in accordance with the CBD. You will find a non exhaustive list of non-monetary and monetary benefits at Appendix II to the Bonn Guidelines: www.biodiv.org/programmes/socio-eco/benefit/bonn.asp;

---

[1] For the purposes of this agreement, commercial application shall mean: applying for, obtaining or transferring intellectual property rights or other tangible or intangible rights by sale or licence or in any other manner; commencement of product development; conducting market research; seeking pre-market approval; and/or the sale of any resulting product.

4. The recipient shall **acknowledge** Kew, as supplier, in all written or electronic reports and publications resulting from their use of the Material, its progeny and derivatives and shall **lodge a copy** of all such publications and reports with Kew;

5. The recipient shall take **all appropriate and necessary measures** to import the Material in accordance with relevant laws and regulations and to contain the Material, its progeny or derivatives so as to prevent the release of invasive alien species;

6. The recipient may only **transfer** the Material, its progeny or derivatives to a bona fide third party such as a botanic garden, university or scientific institution for **non-commercial** use in the areas of scientific research, education, conservation and the development of botanic gardens. All transfers shall be subject to the terms and conditions of this agreement. The recipient shall **notify Kew** of all such transfers and, on request, shall provide Kew with copies of the relevant material transfer agreement;

7. The recipient shall maintain **retrievable records** linking the Material to these terms of acquisition and to any accompanying Data provided by Kew;

8. Unless otherwise indicated, **copyright** in all information or data ("Data") supplied with the Material is owned by Kew or Kew's licensors. You may use this Data on condition that it is used solely for scholarly, education or research purposes; that it is not used for commercial purposes; and that you always acknowledge the source of the Data with the words "With the permission of the Board of Trustees of the Royal Botanic Gardens, Kew";

9. Kew makes **no representation or warranty** of any kind, either express or implied, as to the identity, safety, merchantability or fitness for any particular purpose of the Material, its progeny or derivatives, or as to the accuracy or reliability of any Data supplied. The recipient will indemnify Kew from any and all liability arising from the Material, its progeny or derivatives and/or the Data and from their use or transfer, including any ecological damage. This agreement is governed by and shall be construed in accordance with English law;

10. The recipient will contact Kew to request **prior permission** from Kew or, where appropriate, from the provider of the Material to Kew, for any activities not covered under the terms of this agreement.

I agree to comply with the conditions above:

Signed:	Date: dd/mm/yy
Name and Position:	Organisation and Department:
Address:	E-mail: Tel. Number:

Please return a signed copy to:

....

Royal Botanic Gardens, Kew, Richmond Surrey TW9 3AE, United Kingdom.

Kew Staff Signature:	Name/Position/Date: dd/mm/yy:

LIST OF PLANT MATERIAL SUPPLIED

Appendix 3: South African National Biodiversity Institute Material Supply Agreement

South African National Biodiversity Institute
Agreement for the supply of Biological Material

The South African National Biodiversity Institute (SANBI) is a corporate body constituted under the National Environmental Management Act No. 10 of 2004 and its mission is "to promote the sustainable use, conservation, appreciation and enjoyment of the exceptionally rich plant life of South Africa for the benefit of all its people".

Upon receipt of this Agreement, <u>signed by Recipient below,</u> and because Recipient has agreed to comply with the terms and conditions set forth in the Agreement, the SANBI will supply to Recipient such of the Biological Material[†] requested by recipient as is in the SANBI's sole judgement reasonable and appropriate. Such Biological Material as is supplied to Recipient will be accompanied by a copy of this Agreement, on the reverse of which the Biological Material being supplied (the "Material") will be itemised.

The SANBI intends to honour the letter and spirit of the Convention on Biological Diversity in the use of its collections. Accordingly, the supply of any and all Biological Material by the SANBI to Recipient, including any Material to be supplied under this Agreement, will be subject to the following conditions:

1. Subject to Clauses 2 and 4 below, Recipient may use the Material and any progeny or Derivatives[*] thereof (such as modified or unmodified extracts) for non-commercial purposes only.

2. Recipient will provide the SANBI with a fair and equitable share of any benefits obtained by Recipient arising out of any utilisation by the Recipient of the Material or its progeny or Derivatives, including benefits such as research results and copies of publications. In addition, recipient shall acknowledge the SANBI and, where determinable, the Country of Origin, in all research publications resulting from the use of the Material.

3. **Under this Agreement, Recipient may not Commercialise[#] the Material or any progeny or Derivatives thereof.**

4. If at any point in the future, Recipient wishes to Commercialise the Material or its progeny or derivatives, Recipient must first obtain the written permission of the SANBI. Any Commercialisation to which the SANBI agrees will be subject to a separate agreement between Recipient and the SANBI consistent with the SANBI's policy that benefits should be shared fairly and equitably with the Country of Origin[¥] of the Material

5. **Recipient may not transfer** the Material or any progeny or Derivatives thereof to any party other than recipient or the SANBI without the prior informed consent in writing of the SANBI, and then under a legally binding written agreement, containing terms no less restrictive than those contained in the Agreement unless otherwise agreed in writing by the SANBI.

6. The SANBI makes no representation or warranty of any kind, either express or implied, (1) as to the identity, safety, merchantability or fitness for any particular purpose of the Material or its progeny or Derivatives or (2) that the Material provided to Recipient under this Agreement is or will remain free from any further obligation to obtain prior informed consent from, to share benefits with or to comply with restrictions on use imposed by the country of origin of the Material or any other country or regional economic integration organisation. Recipients will indemnify the SANBI from any and all liability arising out of the Material or its progeny or Derivatives and their use.

7. This Agreement is governed by and shall be construed in accordance with South African law.

I understand that any Material supplied to me by the SANBI pursuant to this Agreement will be subject to, and I agree to comply with, the conditions above.

SIGNED BY: _____

For and on behalf of *[Insert name of recipient institution]* ("**Recipient**")

_____

Name: *[Insert name of individual]* _____

Title: _____

Date: _____

Address of Recipient: _____

_____

SIGNED BY: _____

For and behalf of the NBI

Name: _____

Date: _____

_____

†*Biological material* includes, but is not limited to, plants, plant parts or propagation material (such as seeds, cuttings, roots, bulbs, corms or leaves), fungi or other fungal material, an any other material of plant, animal, fungal, microbial or other origin and the genetic resources contained therein;

Genetic resources mean any material of plant, animal, fungal, microbial or other origin containing functional units of heredity of actual or potential value. This definition of genetic resources is adapted from the definitions of genetic material and genetic resources set forth in Article 2 of the Convention on Biological Diversity;

#*Commercialisation* means the use of or exploitation of genetic resources, their progeny or Derivatives with the object of, or resulting in, financial gain, and includes but is not limited to the following activities: sale, applying for, obtaining or transferring intellectual property rights or other tangible or intangible rights by sale or licence or in any other manner, commencement of product development, conducting market research, and seeking pre-market approval;

¥*Country of Origin* of genetic resources means the country which possesses those genetic resources in *in situ* conditions;

Derivatives include, but are not limited to, modified or unmodified extracts and any compounds or chemical structures based on or derived from genetic resources and their progeny, including analogues.

BIOLOGICAL MATERIAL SUPPLIED:

Appendix 4: American Museum of Natural History Material Transfer Agreement

American Museum of Natural History

Genetic Material Transfer Assurance Form

When requesting tissue samples from the American Museum of Natural History, print, sign and return this form along with your research proposal. The recipient (and the institution which sponsors them) must agree to the conditions specified below. These terms include, but are not limited to, the following:

- Research materials are granted for research purposes only and may not be used in association with human subjects.

- Research materials are not to be used by for-profit recipients for screening, production or sale, for which a commercial license will be required.

- Recipient agrees to comply with all Federal rules and regulations applicable to the research project and the handling of the research material. ***

- The research materials and associated data are provided as a service to the research community. It is being supplied to the recipient with no warranties, express or implied, including any warranty of merchantability or fitness for a particular purpose. Furthermore, AMNH makes no representations that the use of the research material will not infringe any patent or proprietary rights of third parties.

- Unless prohibited by law from doing so, the recipient agrees to hold the AMNH harmless and to indemnify the AMNH for all liabilities, demands, damages, expenses and losses arising out of the recipient's use for any purpose of the research material.

- Research materials must be kept under suitable containment conditions in the lab of the recipient, who agrees not to transfer the research material to other people not under his or her direct supervision without advance written permission from AMNH.

- The recipient is required to acknowledge the AMNH in all presentations or written publications that used data generated from any sample obtained from AMNH, unless requested otherwise. It may also be necessary to acknowledge other individuals, institutions, or their grant numbers for samples donated to the AMNH collection. In any event, the recipient must provide 2 copies of any publications resulting in part or whole from the AMNH tissue grant.

- AMNH requires to be informed of all electronic database submissions (such as GenBank accession numbers) associated with the analytical procedures resulting from the specimens granted. This additional specimen data should be submitted in writing to AMNH.

- AMNH reserves the right to determine the final outcome of any unused portion of the loan material which could include, but is not limited to return of unused material or derivatives thereof to AMNH. This will be reviewed on a case-by-case basis by AMNH and will be specifically addressed in writing on the loan invoice for the materials being transferred. Furthermore, AMNH reserves the right to distribute the research material to others and to use it for its own purposes.

- This Assurance Form shall be construed in accordance with State and Federal laws as applied by the Federal courts and the State of New York.

- Genetic material transfers are made to institutions, not individuals; therefore, this Assurance Form must be signed by an official representative of the borrower's institution. Thus, if you (the researcher) are not an authorized legal signator of your institution, a co-signature from such a representative will also be required.

- Genetic material may not be transferred to another institution without prior written consent of the AMNH.

- By signing this form you are agreeing that you have read and agreed to the terms above.

Date: Signature:

Appendix 5: The Missouri Botanical Garden Material Transfer Agreement for DNA samples

MATERIAL TRANSFER AGREEMENT

The Missouri Botanical Garden releases samples only under specific conditions to support appropriate research projects. Samples in the Garden's DNA Bank have been collected solely for the purpose of supporting molecular phylogenetics and will be released only for the study of relationships of plants or for studies aimed at improving our understanding of evolutionary mechanisms. Samples will not be made available for bioprospecting endeavors, screening for genes of interest in agricultural research, or any other commercial application.

In order to defray a portion of the costs of maintaining, expanding and distributing the special collection, a contribution of $25.00 per sample supplied is requested. For students and those individuals without adequate funding, a request for a complete or partial waiver should be addressed to the Curator of the Herbarium, Missouri Botanical Garden, P.O. Box 299, St. Louis, Missouri, 63166-0299, USA.

As a condition of release for any samples specified on the attached list, each applicant agrees to abide by the restrictions stated above and also agrees to:

1) All requests to pass either material provided by the Garden or extracted DNA to third parties must be approved, via a material transfer agreement, by the Curator of the Herbarium.

2) Acknowledge both the Missouri Botanical Garden and each individual collector of material provided in each publication in which data is used.

3) Provide the Garden with reprints from all resultant publications.

4) Publish jointly with Garden staff members or their foreign collaborators whenever appropriate.

5) Register GenBank/EMBL accession numbers.

Please provide the following information for each request (available on database screen):

Taxa Name: _____

Family: _____

TROPICOS Specimen ID: _____

DNA Sample Type(s): _____

Geography: _____

Collector(s) and number:

_____

Collection Date: _____

I _____(name) of _____ (institutional acronym) certify that I have read and understand the above restrictions and agree that I will conform to all of the regulations of the Missouri Botanical Garden.

Signature _____ Date _____

Please print this form and send the completed version to:

Curator of the Herbarium, Missouri Botanical Garden, P.O. Box 299, St. Louis,

Missouri 63166-0299, USA

Appendix 6: Royal Botanic Gardens, Kew donation letter

**Donation of Material to the
Royal Botanic Gardens, Kew**

The Royal Botanic Gardens, Kew ("Kew") is very grateful to all institutions and individuals who donate material and associated data to Kew in support of Kew's mission of ensuring better management of the Earth's environment by increasing knowledge and understanding of the plant and fungal kingdoms – the basis of life on Earth.

As part of Kew's commitment to upholding the 1992 Convention on Biological Diversity ("CBD") and in accordance with Kew's policy on access and benefit-sharing, before accepting any such material, Kew must be satisfied that material and data has been legally acquired and exported from the country of origin and that Kew is aware of any original terms of acquisition and benefit-sharing. This will enable Kew to ensure that it only accepts material and data into its collections that it can properly and legally curate.

We therefore ask all donors to complete and sign this document. By signing this document the donor confirms that:
- The material and any associated data was, to the best of their knowledge, collected in accordance with all national and international laws and regulations in force at the time of collection: for example, collecting and export permits were obtained and, where required, access and benefit-sharing agreements entered into;
- The donor is aware of, and agrees with, the uses that Kew may make of the material and associated data, once it is accessioned into the Kew collections;
- Copyright in the associated data is hereby assigned to Kew to be used for educational and research purposes, including possible publication in botanical databases on the internet.

1. <u>Material donated:</u> Please list on reverse, or on a separate sheet if more appropriate, the reference numbers and details of herbarium specimens and/or plant or fungal material being donated to Kew.

2. <u>Material legally acquired:</u> Please indicate here if collecting/export permits or access agreements are attached to this document. If documentation is not attached, please explain its absence on reverse, or on a separate sheet if more appropriate.

Documentation Attached/Documentation Not Attached

3. <u>Terms of original acquisition:</u> Please list on reverse, or on a separate sheet if more appropriate, the terms and conditions under which the material or data was originally acquired and any additional restrictions on the future use of the donated material and data.

4. <u>Use of material and data by Kew.</u> Kew may use the donated material and any associated data as follows. It may be:

- Made available for scientific study to Kew staff and authorised visitors; and/or
- Used for the common good in the areas of education and public display; and/or
- Sampled for pollen, spores, DNA, anatomical or cytological preparations and/or chemicals for scientific research purposes; and/or
- Sent on loan, or further distributed, to other scientific institution(s) for further scientific research, on standard terms that prohibit commercialisation unless benefits are shared fairly and equitably with the country of origin of the material; and/or
- Digitally imaged and published in botanical databases freely available on the internet.

Kew will not commercialise material collected after the CBD came into force (29th December 1993) without the prior informed consent of the country of origin and appropriate stakeholders. Kew also undertakes to share fairly and equitably any benefits arising from such commercialisation.

I hereby donate the material and associated data, and assign any copyright in the data, to Kew to use for the stated purposes and declare, that to the best of my knowledge and belief, the information that I have given in this document is accurate.

Name:

Position:

Donating Institution:

Date:

Appendix 7: SANBI/Kew Memorandum of Understanding covering the Darwin Initiative Project: DNA Banking, Phylogeny and Conservation of the South African Flora 2003–2006

NOTES: *The MoU provided below was signed in October 2003, before the South African Biodiversity Act was enacted. It covers particular activities agreed by these two partners – other projects will vary, and the range of potential activities, available funds and the capacities of partners will of course be different. This agreement covers transfer of DNA, not voucher herbarium specimens. For this particular project, such vouchers are kept and curated in the Compton Herbarium at NBI Kirstenbosch.*

MEMORANDUM OF UNDERSTANDING

A Memorandum of Understanding between the National Botanical Institute, South Africa ("NBI") and the Board of Trustees of the Royal Botanic Gardens, Kew ("Kew"), Richmond, Surrey, TWB 3AB, United Kingdom dated this the 7th day of October 2003.

ARTICLE 1: AIMS AND OBJECTIVES

Following many years of inter-institutional links and exchanges, NBI and Kew wish to promote and to safeguard future collaboration, in particular with regard to the creation and maintenance of a DNA bank at the Leslie Hill Molecular Systematics Laboratory (LHMSL) at NBI Kirstenbosch, Cape Town.

NBI and Kew will therefore work together to achieve the objectives of the Darwin Initiative Project: 'DNA Banking, Phylogeny and Conservation of the South African Flora 2003–2006', funded by the United Kingdom's Department for Environment, Food and Rural Affairs (Defra). The details of this project are set out in the 'Project Activities and Time Schedule' document, attached to this Memorandum of Understanding (MoU) as Annex 1.

Kew and NBI will cooperate in the collection and study of DNA from indigenous South African plants (South African DNA) through the establishment of a DNA bank at LHMSL and the establishment of technology training, research and educational programmes.

NBI and Kew further agree that both institutions have endorsed the Principles on Access to Genetic Resources and Benefit-Sharing, and are committed to implementing the 1973 Convention on International Trade in Endangered Species of Wild Fauna and Flora (CITES) and the 1992 Convention on Biological Diversity (CBD) and that, to this end, they will work together to ensure that access to plant material is in accordance with all relevant national laws and regulations and to ensure the fair and equitable sharing of any benefits arising from such access.

ARTICLE 2: INSTITUTIONAL CO-ORDINATORS

RBG Kew:

NBI:

ARTICLE 3: RESPONSIBILITIES – KEW

The following is a list of activities that Kew will pursue in collaboration with NBI in order to promote and achieve the aims of this Darwin Initiative project.

Kew will:

3.1   Contribute to capacity development in the NBI through providing training and assistance in the organisation of any joint Kew/NBI projects such as expeditions and worshops related to this project;

3.2   Assist in the training of the NBI DNA bank manager in DNA extraction techniques and databasing;

3.3   Provide aliquots of Kew's existing DNA bank specimens from South Africa presently housed at Kew. The preparation of the aliquots will be undertaken during the training visit to Kew by the NBI DNA bank manager;

3.4   Provide a copy of the South African portion of Kew DNA bank database to NBI;

3.5   Ensure that the first duplicate sample of the South African DNA collected during any Kew and NBI expedition will be housed at NBI;

3.6   Accession the duplicate DNA sent to Kew into the DNA bank at Kew;

3.7   Maintain the South African DNA in the Kew DNA bank and make it available to Kew staff, students and authorised visitors for research projects to be carried out at Kew;

3.8   Refer all outside requests for South African DNA to the LHMSL;

3.9   Provide a list of the South African DNA that has been studied at Kew to NBI on a twice-yearly basis;

3.10 Inform NBI of any relevant opportunities for training and/or study by appropriate NBI personnel at Kew and elsewhere;

3.11 Coordinate the publication of a Manual on setting up a DNA bank;

3.12 Provide NBI with copies of any project-related publications arising out of its use of the South African DNA, without the need for any request being made;

3.13 Acknowledge the financial support granted by the UK Department for Environment, Food and Rural Affairs (Defra) for this Darwin Initiative project.

ARTICLE 4: RESPONSIBILITIES – NBI

The following is a list of activities that NBI will pursue in collaboration with Kew in order to promote and achieve the aims of this Darwin Initiative project.

NBI will:

4.1 Establish, manage and maintain the DNA bank at LHMSL beyond the term of this project;

4.2 Appoint a DNA bank manager based at the LHMSL at Kirstenbosch;

4.3 Identify and appoint staff to manage and execute this project in accordance with the requirements of South African Labour Legislation;

4.4 Provide educational and training programmes for students from tertiary educational institutions;

4.5 Make existing facilities at LHMSL available to achieve the goals of this project;

4.6 Provide practical help required to facilitate Kew and NBI expeditions and other related project activities;

4.7 Assist researchers from Kew in obtaining all necessary authorisations to enable the lawful collection, study and conservation of the South African DNA, from the relevant authorities and/or other stakeholders. For example, appropriate access and/or export permits, licences and/or prior informed consents;

4.8 Ensure duplicate samples of all South African DNA housed at LHMSL are provided to Kew;

4.9 Allow DNA transferred to Kew to be made available to Kew staff, students and authorised visitors for scientific research purposes;

4.10 Organise project related meetings and training workshops in South Africa;

4.11 Provide conference and workshop facilities, on site accommodation, and office and laboratory space for visiting partners from Kew in relation to this project;

4.12 Write and produce a Field Guide;

4.13 Establish and manage the project website;

4.14 Provide Kew with copies of any project-related publications arising out of its use of the DNA, without the need for any request being made;

4.15 Keep Kew fully informed on all issues relating to implementation of the CBD in South Africa relevant to this project, in particular developments in South Africa's national biodiversity legislation;

4.16 Acknowledge the financial support granted by the UK Department for Environment, Food and Rural Affairs (Defra) for this Darwin Initiative project.

ARTICLE 5: NON-COMMERCIALISATION

5.1 Kew will not Commercialise[2] any South African DNA acquired by Kew under this MoU;

5.2 Without prejudice to the above, any Commercialisation to which Kew and NBI may agree will be subject to a separate written agreement.

ARTICLE 6: OWNERSHIP

Subject to the rights granted by this MoU to Kew to conserve and conduct research upon the South African DNA and enjoy a fair and equitable share of the benefits arising, and subject to the rights of any third party in South Africa, the Government of the Republic of South Africa shall retain ownership of the South African DNA transferred under this MoU.

---

[2] For the purposes of this Memorandum of Understanding, "Commercialisation" shall mean: filing a patent application, obtaining, or transferring intellectual property rights or other tangible or intangible rights by sale or licence or in any other manner; commencement of product development; conducting market research and seeking pre-market approval and/or the sale of any resulting product.

ARTICLE 7: EXCHANGE OF SOUTH AFRICAN DNA

7.1 Transfer of South African DNA between Kew and NBI will be itemised under a notification of transfer in accordance with the terms and conditions of this MoU.

7.2 All South African DNA exchanged between NBI and Kew was acquired and is supplied in accordance with all applicable laws and regulations.

ARTICLE 8: TRANSFER TO THIRD PARTIES

Kew will not supply any of the South African DNA to any third party other than NBI or Kew without prior written authorisation from NBI.

ARTICLE 9: PAYMENT

The annual budget, in pounds sterling, for each financial year will be agreed between Kew and NBI (within the constraints of the budget approved by the Darwin Initiative) and prepared as an addition to Annex 2.

9.1 NBI will submit financial reports and requests for advances in the format provided in Annex 3 within 3 weeks of the end of each financial quarter.

ARTICLE 10: REPORTS

Reports are due on the 31 October and 30 April of each year, starting 31 October 2003. Persons in charge of activities are required to send all necessary materials to the Kew Project Coordinator at least four weeks before the above deadlines.

ARTICLE 11: DURATION AND RENEWAL

11.1 This Memorandum of Understanding will come into force on the date of the second signature. It will be valid for 5 [five] years;

11.2 The obligations and rights contained in Articles 5, 6, 7 and 8 of this MoU shall survive the expiration or other termination of this MoU, unless mutually agreed to the contrary in any replacement agreement.

ARTICLE 12: FORCE MAJEURE

12.1 Neither Kew nor NBI shall be liable to the other party for any delay or non-performance of its obligations under this MoU arising from any cause beyond its reasonable control, including, but not limited to, any of the following: Act of God, governmental act, war, fire, flood, explosion, civil commotion or industrial disputes of a third party.

12.2 The affected party must promptly notify the other party in writing of the cause and the likely duration of the cause. Such notice having been given, the performance of the affected party's obligations, to the extent affected by the cause, shall be suspended during the period the cause persists.

12.3 Without prejudice to the above, the affected party must take all reasonable measures to minimise the impact of any force majeure on the performance of its obligations under the MoU and to ensure, as soon as possible, the resumption of normal performance of the obligations affected by the force majeure.

ARTICLE 13: TERMINATION

This Memorandum of Understanding may be terminated by one party giving three (3) months notice in writing to the other party.

AS WITNESSED IN TWO IDENTICAL COPIES IN THE ENGLISH LANGUAGE, BOTH COPIES BEING EQUALLY AUTHENTIC, BY THE DULY AUTHORISED REPRESENTATIVES OF THE NATIONAL BOTANICAL INSTITUTE AND THE BOARD OF TRUSTEES OF THE ROYAL BOTANIC GARDENS, KEW

SIGNED:	SIGNED:
FOR AND ON BEHALF OF THE NATIONAL BOTANICAL INSTITUTE, SOUTH AFRICA	FOR AND ON BEHALF OF THE BOARD OF TRUSTEES OF THE ROYAL BOTANIC GARDENS, KEW, UNITED KINGDOM
Date:	Date:

Appendix 8: CBD Articles, themes and cross-cutting issues of potential relevance to DNA banks

NOTE: These lists are not exhaustive. For the full CBD text and lists of themes and issues, see www.biodiv.org.

Articles

6 General measures for conservation and sustainable use

7 Identification and monitoring

8 *In situ* conservation

8h Alien species

8j Traditional knowledge

9 *Ex situ* conservation

10 Sustainable use of components of biological diversity

11 Incentive measures

12 Research and training

13 Public education and awareness

14 Impact assessment and minimising adverse impacts

15 Access to genetic resources

16 Access to and transfer of technology

17 Exchange of information

18 Technical and scientific cooperation

19 Handling of biotechnology and distribution of its benefits

Thematic work programmes

Agricultural biodiversity

Dry and sub-humid lands biodiversity

Forest biodiversity

Inland waters biodiversity

Island biodiversity

Marine and coastal biodiversity

Mountain biodiversity

Cross-cutting work programmes

Access to genetic resources

Alien species

Climate change and biological diversity

Economics, trade and incentive measures

Ecosystem approach

Global Strategy for Plant Conservation

Global Taxonomy Initiative

Impact assessment

Indicators

Protected areas

Public education and awareness

Sustainable use of biodiversity

Technology transfer and cooperation

Targets of the Global Strategy for Plant Conservation

1   A widely accessible working list of plant species, as a step towards a complete world flora

2   A preliminary assessment of the conservation status of all known plant species, at a national, regional and international level

3   Development of models with protocols for plant conservation and sustainable use, based on research and practical experience

4   At least 10% of each of the world's ecological regions effectively conserved

5   Protection of 50% of the most important areas for plant diversity assured

6   At least 30% of production lands managed consistent with the conservation of plant diversity

7   60% of the world's threatened species conserved *in situ*

8   60% of threatened plant species in accessible *ex situ* collections, preferably in the country of origin, and 10% of them included in recovery and restoration programmes

9   70% of the genetic diversity of crops and other major socio-economically valuable plant species conserved, and associated local and indigenous knowledge maintained

10   Management plans in place for at least 100 major alien species that threaten plants, plant communities and associated habitats and ecosystems

11   No species of wild flora endangered by international trade

12   30% of plant-based products derived from sources that are sustainably managed

13   The decline of plant resources, and associated local and indigenous knowledge, innovations and practices, that support sustainable livelihoods, local food security and health care, halted

14   The importance of plant diversity and the need for its conservation incorporated into communication, education and public awareness programmes

15   The number of trained people working with appropriate facilities in plant conservation increased, according to national needs, to achieve the targets of this strategy

16   Networks for plant conservation activities established or strengthened at national, regional and international levels

2010 Biodiversity Target

To achieve by 2010 a significant reduction of the current rate of biodiversity loss at the global, regional and national level as a contribution to poverty alleviation and to the benefit of all life on Earth.

The framework includes seven focal areas:

(a) Reducing the rate of loss of the components of biodiversity, including: (i) biomes, habitats and ecosystems; (ii) species and populations; and (iii) genetic diversity;

(b) Promoting sustainable use of biodiversity;

(c) Addressing the major threats to biodiversity, including those arising from invasive alien species, climate change, pollution, and habitat change;

(d) Maintaining ecosystem integrity, and the provision of goods and services provided by biodiversity in ecosystems, in support of human well-being;

(e) Protecting traditional knowledge, innovations and practices;

(f) Ensuring the fair and equitable sharing of benefits arising out of the use of genetic resources; and

(g) Mobilizing financial and technical resources, especially for developing countries, in particular least developed countries and small island developing States among them, and countries with economies in transition, for implementing the Convention and the Strategic Plan.

For a list of the goals, subtargets, and provisional indicators for each of these focal areas, see the CBD website.

LIST OF ABBREVIATIONS

AFLP	amplified fragment length polymorphism
AM-CC	Ambrose Monell Cryo-Collection
AMNH	American Museum of Natural History
APG	Angioperm Phylogeny Group
BRAHMS	Botanical Research and Herbarium Management System
CBD	Convention on Biological Diversity
CFR	Cape Floristic Region
CITES	Convention on International Trade in Endangered Species of Wild Flora and Fauna
CO1	cytochrome oxidase 1
COP	Conference of the Parties
COSHH	control of substances hazardous to health
CsCl	caesium chloride
CTAB	cetyl trimethyl ammonium bromide
DDBJ	DNA Data Bank of Japan
DNA	deoxyribonucleic acid
EDTA	ethylene diamine tetra acetic acid
EMBL	European Molecular Biology Laboratory
EtBr	ethidium bromide
EU	European Union
FAO	Food and Agriculture Organisation
GEF	Global Environment Facility
GPL	General Public License
GPS	global positioning system
GSPC	Global Strategy for Plant Conservation
GTI	Global Taxonomy Initiative
HSS	health, safety and security
IPR	intellectual property rights
IT	information technology
ITPGRFA	International Treaty on Plant Genetic Resources for Food and Agriculture
ITS	internal transcribed spacer
LMO	living modified organism
MBG	Missouri Botanical Garden
MOSAICC	Micro-Organisms Sustainable Use and Access Regulation International Code of Conduct
MoU	Memorandum of Understanding
MSA	material supply agreement

MTA	material transfer agreement
NASA	National Aeronautics and Space Administration
NCBI	National Centre for Biotechnology Information
NFP	national focal point
PAUP	phylogenetic analysis using parsimony
PCR	polymerase chain reaction
PD	phylogenetic diversity
PHP	Hypertext PreProcessor
PHYLIP	Phylogeny Inference Program
PIC	prior informed consent
PPE	personal protection equipment
PVP	poly vinyl pyrrolidone
RAPD	random amplified polymorphic DNA
RB	Rio de Janeiro Botanical Garden Herbarium
rbcL	ribulose 1,5-bisphosphate carboxylase large subunit
RNA	ribonucleic acid
SANBI	South African National Biodiversity Institute
SEVAG	chloroform:isoamyl alcohol 24:1
Tris HCl	tris (hydroxymethyl) aminomethane hydrochloride
UNEP	United Nations Environment Programme
URL	uniform resource locator
UV	ultraviolet

SELECTED REFERENCES

Section A: The DNA molecule and its uses in biodiversity and conservation

Agapow, P.M., O.R.P. Bininda-Emonds, K.A. Crandall, J.L. Gittleman, G.M. Mace, J.C. Marshall and A. Purvis. 2004. The impact of species concept on biodiversity studies. *Quarterly Review of Biology* 79: 161–179.

Donoghue, P.C.J. and M.P. Smith (eds.). 2003. *Telling the evolutionary time: molecular clocks and the fossil record.* Taylor and Francis, London.

Doolittle, W.F. and C. Sapienza. 1980. Selfish genes, the phenotype paradigm and genome evolution. *Nature* 284: 601–603.

Faith, D.P. 1992. Conservation evaluation and phylogenetic diversity. *Biological Conservation* 61: 1–10.

Felsenstein, J. 2004. *Inferring phylogenies.* Sinauer Associates Inc, Sunderland.

Hillis, D.M., C. Moritz and B.K. Mable (eds.). 1996. *Molecular systematics,* 2nd edition. Sinauer Associates Inc, Sunderland.

Mace, G.M., J.L. Gittleman and A. Purvis. 2003. Preserving the tree of life. *Science* 300: 1707–1709.

Orgel, L.E. and F.H.C. Crick. 1980. Selfish DNA – the ultimate parasite. *Nature* 284: 604–607.

Page, R.D.M. and E.C. Holmes. 1998. *Molecular evolution, a phylogenetic approach.* Blackwell Science, Oxford.

Pagel, M. 1999. Inferring the historical patterns of biological evolution. *Nature* 401: 877–884.

Roca, A.L., N. Georgiadis, J. Slattery-Pecon and S. J. O'Brien. 2001. Genetic evidence for two species of elephants in Africa. *Science* 293: 1473–1477.

Savolainen, V. and M.W. Chase. 2003. A decade of progress in plant molecular phylogenies. *Trends in Genetics* 19: 717–724.

Sugden, A.M., B.R. Jasny, E. Culotta and E. Pennisi. 2003. Charting the evolutionary history of life. *Science* 300: 1691. (Special section on the Tree of Life).

Tautz, D., P. Arctander, A. Minelli, R.H. Thomas and A.P. Vogler. 2002. DNA points the way ahead in taxonomy. *Nature* 418: 479.

Tautz, D., P. Arctander, A. Minelli, R.H. Thomas and A.P. Vogler. 2003. A plea for DNA taxonomy. *Trends in Ecology and Evolution* 18: 70–74.

Thomas, C.D., A. Cameron, R.E. Green, M. Bakkenes, L.J. Beaumon, Y.C. Collingham, B.F.N. Erasmus, M.F. de Siqueira, A. Grainger, L. Hannah, L. Hughes, B. Huntley, A.S. van Jaarsveld, G.F. Midgley, L. Miles, M.A Ortega-Huerta, A.T. Peterson, O.L. Phillips and S.E. Williams. 2004. Extinction risk from climate change. *Nature* 427: 145–148

Watson, D.J. and F.H.C. Crick. 1953. Molecular structure of nucleic acids. *Nature* 171: 737–738.

Section B: Legal issues surrounding DNA banking

Bragdon, S. (ed.). 2004. International Law of Relevance to Plant Genetic Resources: A practical review for scientists and other professionals working with plant genetic resources. *Issues in Genetic Resources* no. 10, March 2004. International Plant Genetic Resources Institute, Rome.

Cheyne, P. 2004. Access and benefit-sharing agreements: bridging the gap between scientific partnerships and the Convention on Biological Diversity. Pp. 3–26. In: R.D. Smith, J.B. Dickie, S.H. Linington, H.W. Pritchard and R.J. Probert (eds.) *Seed conservation: turning science into practice.* Royal Botanic Gardens, Kew.

CITES. 2003 and updates. *CITES Handbook.* Secretariat of the Convention on International Trade in Endangered Species of Wild Fauna and Flora, Geneva.

Glowka, L., F. Burhenne-Guilmin and H. Synge in collaboration with J. McNeely and L. Gündling. 1994. *A guide to the convention on biological diversity.* IUCN, Gland and Cambridge.

Laird, S.A. (ed.). 2002. *Biodiversity and traditional knowledge: equitable partnerships in practice.* People and Plants Conservation Series, Earthscan Publications Ltd., London.

Latorre, F., C. Williams, K. ten Kate and P. Cheyne. 2002. *Results of the pilot project for botanic gardens: principles on access to genetic resources, common policy guidelines to assist with their implementation and explanatory text.* Royal Botanic Gardens, Kew. Also available at www.kew.org/conservation.

Mackenzie, R., F. Burhenne-Guilmin, A.G.M. La Viña and A. Werksman, in cooperation with A. Ascencio, J. Kinderlerer, K. Kummer and R. Tapper. 2003. An explanatory guide to the cartagena protocol on biosafety. *IUCN Environmental Policy and Law Paper* no. 46. IUCN, Gland and Cambridge.

McGough, H.N., M. Groves, M. Mustard and C. Brodie. 2004. *CITES and plants: a user's guide.* Royal Botanic Gardens, Kew.

Paton A, C. Williams and K. Davis. 2005. Taxonomy in the implementation of the Convention on Biological Diversity. Pp. 18-28. In: E. Leadley & Jury, S. (eds.) *Taxonomy and plant conservation.* Cambridge University Press, Cambridge.

Pew Conservation Scholars Initiative. 1995. Suggested ethical guidelines for accessing and exploring biodiversity. *Eubios Journal of Asian and International Bioethics* 5:38-40. Also available at www.csu.edu.au/learning/eubios/EJ52I.html.

Williams, C., K. Davis and P. Cheyne. 2003. *The CBD for botanists: an introduction to the Convention on Biological Diversity for people working with botanical collections.* Royal Botanic Gardens, Kew.

Section C: Practical considerations for DNA and tissue banking

Bridson, D. and L. Forman. 1992. *The Herbarium Handbook*, revised edition. Royal Botanic Gardens, Kew.

Chase, M.W. and H.H. Hills. 1991. Silica gel: an ideal material for field preservation of leaf samples for DNA studies. *Taxon* 40: 215–220.

Dessauer, H.C. and M.S. Hafner. 1984. *Collections of frozen tissues: value, management, field and laboratory procedures and directory of existing collections.* Association of Systematics Collections, University of Kansas Press, Lawrence, Kansas.

Doyle, J.J and J.L. Doyle. 1987. A rapid DNA isolation procedure for small quantities of fresh leaf tissue. *Phytochemical Bulletin* 19: 11–15.

HSC Health & Safety Commission. 1989. *Control of substances hazardous to health (General ACoP and control of carcinogenic substances (Carcinogens ACoP)).* Her Majesty's Stationery Office, London.

HSE Health & Safety Executive. 1989. *COSHH Assessments: Control of Substances Hazardous to Health Regulations 1988.* Her Majesty's Stationery Office, London.

Harris, R.H. 1990. Zoological preservation and conservation techniques. *Journal of Biological Conservation* 1: 4–68.

Huber, J.T. 1998. The importance of voucher specimens, with practical guidelines for preserving specimens of the major invertebrate phyla for identification. *Journal of Natural History* 32: 367–385.

Lee, W.L., B.M. Bell and J.F. Sutton. 1982. *Guidelines for acquisition and management of biological specimens. A report of the participants of a conference on voucher specimen management sponsored under the auspices of the council of curatorial methods of the association of systematics collections.* Association of Systematics Collections, University of Kansas Press, Lawrence, Kansas.

Monk, R.R. and R.J. Baker. 2001. E-vouchers and the use of digital imagery in natural history collections. *Museology* 10: 1–8.

Meester, J. 1990. The importance of retaining voucher specimens. Pp. 123–127. In: E.M. Herholdt (ed): *Natural history collections: their management and value*. Transvaal Museum Special Publication no. 1, Transvaal Museum, Pretoria.

Palmer, J.D., R.K. Jansen, H.J. Michaels, M.W. Chase and J.R. Manhart. 1989. Chloroplast DNA variation and plant phylogeny. *Annals of the Missouri Botanic Garden* 75: 1180–1206.

Prendini, L., R. Hanner and R. DeSalle. 2002. Obtaining, storing and archiving specimens for molecular genetic research. Pp. 176–248. In: R. DeSalle, G. Giribet and W. Wheeler (eds.) *Methods and tools in biosciences and medicine (MTBM) techniques in molecular systematics and evolution*. Birkhauser Verlag, Basel.

Royal Society of Chemistry. 1996. *COSHH in laboratories*. The Safe Laboratory Series, Royal Society of Chemistry, Cambridge.

Saghai-Maroof, M.A., K.M. Soliman, R.A. Jorgensen and R.W. Allard. 1984. Ribosomal DNA spacer-length polymorphisms in barley: Mendelian inheritance, chromosomal location, and population dynamics. *Proceedings of the National Academy of Sciences USA* 81: 8014–8018.

Section D: Case studies

Adams, R.P.N.Do and C. Ge-lin. 1992. Preservation of DNA in plant specimens from tropical species by desiccation. Pp. 135-152. In: R. P. Adams and J.E. Adams (eds.) *Conservation of plant genes: DNA banking and in vitro biotechnology*. Academic Press, San Diego.

Baker, R. J. 1994. Some thoughts on conservation, biodiversity, museums, molecular characters, systematics and basic research. *Journal of Mammalogy* 75: 277–287.

Barrett, S.C.H. and J.R. Kohn. 1991. Genetic and evolutionary consequences of small population size in plants: implications for conservation. Pp. 1–30. In: D.A. Falk and K.E. Holsinger (eds.) *Genetics and conservation of rare plants*. Oxford University Press, New York.

Bowles, M.L. and C.J. Whelan (eds.). 1994. *Restoration of endangered species: conceptual issues, planning and implementation*. Cambridge University Press, Cambridge.

Brown, A.H.D. and J.D. Briggs. 1991. Sampling strategies for genetic variation in ex situ collections of endangered plant species. Pp. 99–122. In: D.A. Falk and K.E. Holsinger (eds.) *Genetics and conservation of rare plants*. Oxford University Press, New York.

Coates, D.J. and V.L. Hamley. 1999. Genetic divergence and mating system in the endangered and geographically restricted species, *Lambertia orbifolia* Gardner (Proteaceae). *Heredity* 83: 418–427.

Cochrane, A. 2004. Western Australia's ex situ program for threatened species: a model integrated strategy for conservation. Pp. 41–66. In: E.O. Guerrant Jr., K. Havens and M. Maunder (eds.) *Ex situ plant conservation: supporting species survival in the wild.* Island Press, Washington.

Cook, J.A., G.H. Jarrell, A.M. Runck and J.R. Demboski. 1999. *The Alaska frozen tissue collection and associated electronic database: a resource for marine biotechnology.* OCS Study MMS 99-0008. University of Alaska, Fairbanks.

Dessauer, H.C. and H.S. Hafner. 1984. *Collections of frozen tissues: value, management, field and laboratory procedures, and directory of existing collections.* Association of Systematics Collections, Washington, D.C.

Dixon, K.W. 1994. Towards integrated conservation of Australian endangered plants – the Western Australian model. *Biodiversity and Conservation* 3: 148-159.

Engstrom, M.D., R.W. Murphy and O. Haddrath. 1999. Sampling vertebrate collections for molecular research: practice and policies. Pp. 315-330. In: D. Metsger and S. Byers (eds.). *Managing the modern herbarium: an interdisciplinary approach.* Elton-Wolf, Vancouver.

Falk, D.A., C.J. Millar and M. Olwell (eds.). 1996. *Restoring diversity: strategies for reintroduction of endangered plants.* Island Press, Washington D.C.

Fay, M.F. and R.S. Cowan. 2001. Plastid microsatellites in *Cypripedium calceolus* (Orchidaceae): genetic fingerprints from herbarium specimens. *Lindleyana* 16: 151-156.

Fay, M.F., R.S. Cowan and I.J. Leitch. 2005. The effects of genome size on the quality and utility of AFLP fingerprints. *Annals of Botany* 95: 237-246.

Fay, M.F. and S.L. Krauss. 2003. Orchid conservation genetics in the molecular age. Pp. 91-112. In: K.W. Dixon, S.P. Kell, R.L. Barrett and P.J. Cribb (eds.). *Orchid Conservation.* Natural History Publications (Borneo), Kota Kinabalu.

Fiedler, P.L. and P.M. Kareiva (eds.). 1998. *Conservation biology for the coming decade*, 2nd edition. Chapman and Hall, New York.

Florian, M.L. 1990. The effects of freezing and freeze-drying on natural history specimens. *Collection Forum* 6: 45-52.

Franks, F. 1985. *Biophysics and biochemistry at low temperatures.* Cambridge University Press, Cambridge.

Goldblatt, P. 1997. Floristic diversity in the Cape flora of South Africa. *Biodiversity and Conservation* 6: 359–377.

Goldblatt, P. and J.C. Manning. 2002. Plant diversity of the Cape region on southern Africa. *Annals of the Missouri Botanical Garden* 89: 281-302.

Guerrant Jr., E.O., K. Havens and M. Maunder (eds.). 2004. *Ex situ plant conservation: supporting species survival in the wild.* Island Press, Washington.

Higuchi, R., B. Bowman, M. Freiberger, O.A. Ryder and A.C. Wilson. 1984. DNA sequences from the Quagga, an extinct member of the horse family. *Nature* 312: 282-284.

Husband, B.C. and L.G. Campbell. 2004. Population responses to novel environments: implications for ex situ plant conservation. Pp. 231–266. In: E.O. Guerrant Jr., K. Havens and M. Maunder (eds.) *Ex situ plant conservation supporting species survival in the wild.* Island Press, Washington, D.C.

Ioannou, Y.A. 2000. A frozen database. *Science* 288: 1191.

Jansen, R.K., D.J. Loockerman and H.-G. Kim. 1999. DNA sampling from herbarium material: a current perspective. Pp. 277-286. In: D.M. Metzger and S.C. Byers (eds.) *Managing the modern herbarium: an interdisciplinary approach.* Society for the Preservation of Natural History Collections, Washington, D.C.

Krauss, S.L., B. Dixon and K.W. Dixon. 2002. Rapid genetic decline in a translocated population of the rare and endangered *Grevillea scapigera* (Proteaceae). *Conservation Biology* 16: 986-994.

Linder, H.P. 2003. The radiation of the Cape flora, southern Africa. *Biological Review* 78: 597–638.

Maunder, M., A. Culham and C. Hankamer. 1998. Picking up the pieces: botanical conservation on degraded oceanic islands. Pp. 317–344. In: P.L. Fiedler and P.M. Kareiva (eds.) *Conservation biology for the coming decade*, 2nd Edition. International Thomson Publishing, New York.

Mistretta, O. 1994. Genetics of species re-introductions: applications of genetic analysis. *Biodiversity and Conservation* 3: 184–190.

Mueller, G.M. 1999. A new challenge for mycological herbaria: destructive sampling of specimens for molecular data. Pp. 287-300. In: D.M. Metzger and S.C. Byers (eds.) *Managing the modern herbarium: an interdisciplinary approach.* Society for the Preservation of Natural History Collections, Washington, DC.

Rieseberg, L.H. and S.M. Swensen. 1996. Conservation genetics of endangered island plants. Pp. 305–334. In: J.C. Avise and J.L. Hamrick (eds.) *Conservation genetics: case histories from nature.* Chapman and Hall, New York.

Ruedas, L.A., J. Salazar-Bravo, J.W. Dragoo and T.L. Yates. 2000. The importance of being earnest: what, if anything, constitutes a specimen examined. *Molecular Phylogenetics and Evolution* 17: 129–132.

Scherczinger, C.A., C. Ladd, M.T. Bourke, M.S. Adamowicz, P.M. Johannes, R. Scherczinger, T. Beesley and H.C. Lee. 1999. A systematic analysis of PCR contamination. *Journal of Forensic Sciences* 44: 1042–45.

Thomas, R.H. 1994. Molecules, museums and vouchers. *Trends in Ecology and Evolution* 9: 413–414.

Thomas, W.K., S. Pääbo, F.X. Villablanca and A.C. Wilson. 1990. Spatial and temporal continuity of kangaroo rat population shown by sequencing mitochondrial DNA from museum specimens. *Journal of Molecular Evolution* 31: 101–112.

Turner, S., S.L. Krauss, E. Bunn, T. Senaratna, K.W. Dixon, B. Tan and D. Touchell. 2001. Genetic fidelity and viability of *Anigozanthos viridis* following tissue culture, cold storage and cryopreservation. *Plant Science* 161: 1099–1106.

Van Wyk, E. and G.F. Smith. 2001. *Regions of floristic endemism in southern Africa.* Umdaus Press: Hatfield.

Vos, P., R. Hogers, M. Bleeker, M. Rijans, T. Van de Lee, M. Hornes, A. Frijters, J. Pot, M. Kuiper, M. Zabeau 1995. AFLP: a new technique for DNA fingerprinting. *Nucleic Acids Research* 23: 4407–4414.

Wood, E.W., T. Eriksson and M.J. Donaghue.1999. Guidelines for the use of herbarium materials in molecular research. Pp. 265–276. In: D.M. Metzger and S.C. Byers (eds.) *Managing the modern herbarium: an interdisciplinary approach.* Society for the Preservation of Natural History Collections, Washington, D.C.

INDEX

access to genetic resources 18–26, 28, 36–46, 76–7

acquisition of material 33, 36–40, 44–6, 66–70, 129–36

agreements 21, 38–41, 44–6
  access and benefit-sharing 44, 46, 129
  benefit-sharing 38, 44
  MSA 44, 104, 118–124
  MTA 24, 26, 28–9, 37–9, 41, 44–6, 61, 72, 77, 85, 119, 125–8
  MoU 38, 44–6, 79, 131–6

Ambrose Monell Cryo-Collection (AMCC) 107–113, 125–6

American Museum of Natural History 70, 107–8, 125–6

benefit-sharing 18–26, 28–9, 36–9, 41–6, 76–7, 79, 118, 122, 129–30, 132–3, 137, 140

Biodiversity Act no.10 (South Africa) 76–8, 131, 121

bioprospecting 3, 75–7, 127

Bonn Guidelines 21, 24–6, 46, 118

Cartagena Protocol on Biosafety 20, 28

CBD *see* Convention on Biological Diversity

CITES *see* Convention on International Trade in Endangered Species of Wild Fauna and Flora

clonality 93–4

codes of conduct 21, 25–6

collection techniques 66–8, 83–5, 101

commercialisation 25, 29, 36, 42–3, 46, 75–7

conservation genetics 80, 87–95, 100–6

Convention on Biological Diversity (CBD) 18–28, 36–46, 64, 72, 76, 79–80, 87, 99, 118, 121, 129, 132, 134
  Article 15: 22–3, 36–7
  GSPC 18, 21, 27, 78, 87, 138–9
  GTI 18, 21, 26–7

National Focal Points 19, 27, 38, 45
  objectives 18
  practical implementation 36–46

Convention on International Trade in Endangered Species of Wild Fauna and Flora (CITES) 30–40, 44–5, 60, 132
  exemptions (for registered scientific institutions) 31, 33–5, 39
  permits 30–1, 33–5, 38–9, 44, 60
  simplified procedures 35
  stricter domestic measures 30–1, 33, 37

cryostorage 87, 93, 107–13

cryptic species 10–12, 27

Darwin Initiative 46, 72, 79, 132

databasing 40, 61–5, 111–13

DDBJ 41

derivatives 20, 31, 76–7, 124

DNA
  barcodes 4, 10, 12, 14–16, 27
  extraction 46–60, 101, 114–7
  fingerprinting 4, 16, 89, 91, 103
  markers *see* markers
  sequencing 2–3, 6, 105
  storage 57, 84, 98, 104, 108, 110–12, 116
  structure 3

donation letter 39, 129–30

drying techniques 66

EMBL 41

ethidium bromide 49–51, 53, 55–7, 58–9, 115–7

e-voucher *see* vouchers

forensics 12–16, 81

GenBank 41, 61–3, 112

genetic erosion 91–2

genetic fidelity 89, 91, 93

genetic resources 18, 20–6, 28, 37, 43, 72, 76–8, 96, 123